DIY Hydroponics Gardening

A 2-in-1 beginner's Guide to Growing Fruits and Vegetables in Your Own Organic Greenhouse Garden All Year-Round. Learn Easy & Inexpensive Hydroponic & Aquaponic Techniques

Scott Fields

2 Books in 1:

o *DIY Hydroponics*
o *Hydroponics Gardening*

Copyright 2020 - All rights reserved.

The content contained within this book may not be reproduced, duplicated or transmitted without direct written permission from the author or the publisher.

Under no circumstances will any blame or legal responsibility be held against the publisher, or author, for any damages, reparation, or monetary loss due to the information contained within this book, either directly or indirectly.

Legal Notice:

This book is copyright protected. It is only for personal use. You cannot amend, distribute, sell, use, quote or paraphrase any part, or the content within this book, without the consent of the author or publisher.

Disclaimer Notice:

Please note the information contained within this document is for educational and entertainment purposes only. All effort has been executed to present accurate, up to date, reliable, complete information. No warranties of any kind are declared or implied. Readers acknowledge that the author is not engaging in the rendering of legal, financial, medical or professional advice. The content within this book has been derived from various sources. Please consult a licensed professional before attempting any techniques outlined in this book.

By reading this document, the reader agrees that under no circumstances is the author responsible for any losses, direct or indirect, that are incurred as a result of the use of information contained within this document, including, but not limited to, errors, omissions, or inaccuracies.

DIY Hydroponics

The Beginner's Guide to Building a Sustainable and Inexpensive Hydroponic System at Home. Learn How to Quickly Start Growing Plants in Water.

Scott Fields

Table of Contents

Introduction 1

Chapter 1: Origin of Hydroponics 5
   Hydroponics Farming 7

Chapter 2: Different Types of Hydroponics and Hydroponics Systems 13
   The Drip System or Continuous Drip 14
   The Ebb and Flow System 16
   The Water Culture Growing System 19
   Falling Water 21
   Recirculating Water 22
   The Wick System 23

Chapter 3: Plant Nutrition 27
   Homemade Nutrients 27
   USE OF NUTRIENT SOLUTIONS 36

Chapter 4: Actual Plant Growth 39

Chapter 5: Outdoor vs. Indoor 47
   Hydroponics: Comparing Indoor and Outdoor Growing 47
   Lighting 47

Chapter 6: Inexpensive Hydroponics 55
   Why hydroponics? 60
   Warnings 66
   Hydroponic Garden Backyard vs. Hydroponic Greenhouse 66

Chapter 7: Which Specific Materials/Equipment Do You Need?..........69
    PH Meter 70
    EC Meter 71
    TDS Meter 73
    Dissolved Oxygen Sensor 75

Humidity and temperature sensor 76

Germination Tray and Dome 77

Seed Starter cubes 77
    Rockwool Cubes 78
    Coco Coir 78
    Oasis Cubes 79
    Sponges 79
    Hydration 80
    Perlite 82
    Vermiculite 83
    Rockwool 83
    Grow Stones 84
    River Rock 84
    Pine Shavings 85

Chapter 8: How to Build Your Hydroponic System 87

Steps Involved 87

Tools and Everything You Need 88

Growing Chamber 90

Delivery System 91

Submersible Pump 92

Air Pump 92

Timer 93

Growing Medium 94

Growing Lights 95

pH Testing Kit 97

The Nutrient Solution 98

Chapter 9: Growing Mediums, Nutrients, and Lights *99*

Light 99

High-Intensity Discharge (HID) Lighting 99

Intensity 101

Duration (Photoperiod) 102

Light (Photosynthetic spectrum) 102

Choosing a Grow Light 103

Growing Plants 104

Temperature 105

Humidity 105

Light 105

Co2 106

Dissolved Oxygen 106

pH 107

Total Dissolved Solids (TDS) 107

Grower's Guide to Common Plants 107

    Important Note: 108

Getting Started with Seeds 108

Seed Starting 110

Chapter 10: Advantages and Disadvantages of Hydroponics *113*

No Soil Required 114

Make Better Use of Space and Location 114

Climate Control 115

Hydroponics Saves Water 115

Efficient Use of Nutrients 116

pH Control of the Solution 116

No Weeds 116

Fewer pests and diseases 116

Reduced Need for Insecticides, Herbicides, and Other Chemicals 117

Work and Time Saving 117

Hydroponics is a hobby that relieves stress 117

Cons and Challenges 118

A hydroponic garden requires time and commitment 118

Chapter 11: New Techniques in Hydroponics 123

Hydroponics and quality of vegetable products........................... 123

Description of the hydroponic system 125

Bioactive compounds 127

Example of health benefits of chemical classes 127

Hydroponics and accumulation of bioactive compounds............ 129

Crop Results Reference 131

Chapter 12: Maintenance of Your Hydroponic System... 137

Required Daily Hydroponic System Maintenance 137

Regular Hydroponic System Maintenance Check List................ 138

Learn to identify the good and bad bugs. 140

Required bi-monthly maintenance 140

Chapter 13: Potential Problems and Solutions 145

Pests 145
Organic Pesticides 146
Leaves 147
Algae Growth 148
System Maintenance 149
The Lighting 149
   Consider These Things When Using Natural Lighting 149
   Artificial Lighting 150
The Growing Climate 150
The Nutrient System 151

Conclusion *153*

Introduction

No Soil
Lorem ipsum dolor sit amet, consectetur adipiscing elit, sed do eiusmod tempor

Fast Growth
Lorem ipsum dolor sit amet, consectetur adipiscing elit, sed do eiusmod tempor

Less diseases
Lorem ipsum dolor sit amet, consectetur adipiscing elit, sed do eiusmod tempor

Less pesticide use
Lorem ipsum dolor sit amet, consectetur adipiscing elit, sed do eiusmod tempor

Water-Saving
Lorem ipsum dolor sit amet, consectetur adipiscing elit, sed do eiusmod tempor

Higher Yields
Lorem ipsum dolor sit amet, consectetur adipiscing elit, sed do eiusmod tempor

Affordable
Lorem ipsum dolor sit amet, consectetur adipiscing elit, sed do eiusmod tempor

Hydroponics is the study of soilless plants. You might have heard of soilless society, which is another term to explain hydroponics. We use the same natural elements to grow plants in soils, so that weeds, soil-borne pests, and diseases do not harm the plants.

Once you develop a plant, its production is higher than average, whether you grow it in a greenhouse, a garden, or on a balcony. Hydroponics also helps you have more plants per square meter. Since the plants don't have to compete with weeds, the food and water is supplied directly. Despite many myths, plants grown in hydroponics are similar to plants grown in the soil. Plants cultivated in a hydroponic system require the same nutrients as those produced in the land, but you can regulate the quality more precisely. The fundamental difference between the two is how you supply the plants with nutrients and water.

We already process nutrient salts in hydroponics, so plants do not have to wait before nutrients fall into the necessary form. For soil agriculture, however, plants are fed nutrients through manure and compost, which must be broken down into their basic shape (nutrient salts) before the plants can use them.

Hydroponic systems often use artificial lighting. This can significantly increase the cost of a program. If you have pure sunlight, artificial lighting will not be necessary. If you do not, the initial costs may be relatively higher; many of our customers find that a lighting system can be a problem and is an ongoing expense that they did not expect.

Some people are nervous about the hydroponics concept, mostly because of the use of various technology components. Yet, hydroponics is a simple and easy method of growing plants that is no more difficult than plant cultivation in soil. We at Aqua Gardening practice hydroponics and are more than prepared to introduce you to this innovative technology so that you, too, can grow fresh and healthy plants.

For decades, hydroponic techniques have been in use. The earliest hydroponics use was in Babylonian Hanging Gardens, Kashmir Floating Gardens, and the Mexican Aztecs, who have grown plants using rafts on low-lying lakes. Furthermore, hieroglyphic texts dating back to ancient Egypt show references to hydroponics. Most recently, mobile hydroponic farms were used to feed the troops in the South Pacific during World War II.

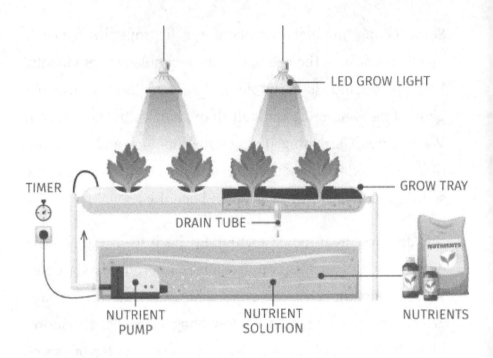

Hydroponics is now playing a more significant role in agricultural production around the world and has recently become more accessible for the household grower. Demand for environmentally friendly and safer goods in society has been an essential factor in this development. You will know precisely what has happened to the plants by growing them in a hydroponic system and can guarantee no use of harmful chemicals, which can affect your health and the ecosystem.

Chapter 1: Origin of Hydroponics

Hydroponics is a Greek term coming from two words - "Hydro," which means "Water" and "Ponos," meaning "labor." It translates to "working water." Hydroponics is the soilless growth of plants. It is a subset of hydroculture, which involves the growing of plants in a water-based, nutrient-rich solution. Instead of soil, the root system gets support from an inert medium like vermiculite, perlite, clay pellets, Rockwool, or peat moss. In this practice, plant roots come in direct contact with the nutrient solution while having access to oxygen, which is vital for plant growth. The nutrients come from sources such as duck manure, fish excrement, or purchased organic fertilizers. Common examples of plants that are grown hydroponically are:

   Tomatoes
   Peppers
   Cucumbers
   Lettuces
   Marijuana

The Hanging Gardens of Babylon (600BC) located along the River Euphrates in Babylonia and the floating gardens in China (in the late 1200s) are some of the earliest manifestations of the hydroponic culture.

The first academic work discussing growing plants without soil was published in 1627, in a book called, "A Natural History" by Francis Bacon. It sparked a shift in conventional beliefs and triggered an increase in the research of the water culture technique. In 1699, an Englishman named John Woodward used water that contained soil mixes to grow plants. He concluded that plants get their nutrients from certain minerals and substances from the water in the soil.

Two German botanists, Julius von Sachs and Wilhelm Knop, developed a soilless plant cultivation technique from 1859-1875. The soilless growth of terrestrial plants in mineral nutrients solutions was called "solution culture." It became a standard research area and the technique is still widely used and taught.

William Frederick Gericke, at the University of California at Berkeley, began public enlightenment and activism campaigns for growing plants through solution culture. He called it "aquaculture" at first but found out that it more accurately applies to the culture of aquatic organisms. He introduced the term "Hydroponics" in 1937.

One of the earliest known successes of hydroponics was in the 1930s in a place called Wake Island, a rocky island located in the Pacific Ocean used by Pan American Airlines as a refueling station. They used this technique to grow vegetables for passengers because there was no soil, and it was expensive to transport fresh vegetables. Countries like Italy, France, Spain, Israel, Germany, England, and USSR, among others, started hydroponics in commercial farms and greenhouses. NASA has done numerous researches in hydroponics for its Controlled Ecological Life Support System (CELSS). In 2007, Eurofresh Farms located in Wilcox, Arizona, sold over 200 million pounds of tomatoes grown hydroponically.

As of 2017, several hydroponic greenhouses produce tomatoes, cucumbers, and peppers. Technological advancements in this field alongside other economic factors have led to projected growth of $724.87 billion by the year 2023.

Hydroponics Farming

1. In hydroponics, you do not need soil to grow your plants. It invariably means that plants can be grown in areas where the soil is limited, non-existent, or heavily contaminated. NASA is looking into hydroponics to grow plants for astronauts in space.

2. Hydroponics helps you to maximize space. Everything the plants need to grow are obtained from and sustained in a system, and as such, plants can be cultivated in any spare space. As plants grow, their roots expand and the search for oxygen and food increases. In hydroponics, the roots are

placed in a chamber that contains oxygenated nutrient solutions and come in direct contact with essential minerals.

3. Like greenhouses, you can grow plants at any time of the year. You have control over light intensification, temperature, humidity, and air composition, among other important variables. Commercial farmers can exploit this feature by growing plants at specific times to boost profits. Harvesting under hydroponics is also very easy as well.

4. Hydroponics saves 70-90% more water than soil because the water is recirculated and reused. Plants use the amount of water they need while the run-offs become trapped, taken back, and reused by the system. There is limited water loss as it only occurs by two methods—via evaporation or any leakages from the system. However, a good hydroponic setup minimizes or eliminates any potential leaks. Agriculture uses an estimated 80% of ground and surface water in the United States. As food production increases in the future, water is sure to become a critical issue. Hydroponics saves us from these problems.

5. Plants grown hydroponically have a better growth rate. You are literally in control of almost every essential factor in the growth of the plants. The plants are under ideal conditions, nutrients are provided in the right amounts, and the plant is in direct contact with the root system. Energies

that plants would have wasted will be conserved and used for developing and making healthy fruits.

6. You do not have to worry about weeds. Those little shrubs are irritants to every farmer. Weed control is one of the greatest tasks that stresses gardeners and consumes time. We know that weeds are related to soil; hence, once we eliminate soil, we invariably eliminate weeds. You also do not have to worry too much about pests or diseases. The absence of soil means the absence of pests that are inherent in the soil as well. Groundhogs, birds, gophers, and diseases like Pythium and Fusarium will be a thing of the past.

7. Hydroponics is a good way to relax and relieve yourself of work stress. It is one of the easiest ways of being in touch with nature. You are spared the "I don't have gardening space" excuse and given a healthy hobby. You can start with tasty fresh vegetables or vital herbs. The sight of fresh greens in your room will surely revitalize your spirit.

8. There is a lot of stress with traditional farming. You have to till the land with the aid of machines or with your hands. You move around the farm, diligently planting your seeds. You must dedicate a huge chunk of your time weeding, watering, and taking care of your plants. All of these are eliminated in hydroponics. You get to farm more easily.

9. In traditional farming methods, you are always in need of a water source for your plants. Sometimes, this leads to an unwarranted increase in monies to be spent taking care of the farm. In hydroponics, the nutrients are placed in water that can be recirculated. It allows you to conserve water and saves you the stress of watering your plants daily. It also reduces costs, as you only have to change the nutrient solution after a few days.

10. Controlling pH levels is sometimes not given the attention it deserves by growers, despite its importance to the cultivation process. It gives your plants the freedom to access the required quantities of nutrients that are needed for their growth and development. Unlike traditional farming, hydroponics provides every important nutrient the plant needs in the growing solution. The pH of this solution could be altered and correctly measured so that a correct and optimal pH level is maintained every time. When the pH is at an optimal level, plants uptake nutrients faster and better. If the pH level varies excessively, plants will suffer because they cannot absorb nutrients. Optimal pH levels range from 5.5-7, though some plants do extremely well in an environment that is slightly acidic. It will be good to research optimal pH levels for the plants you are about to grow and how it can be effectively regulated under a hydroponic environment.

11. With hydroponics, you do not have to worry about the weather. Whether it is growing tomatoes on your windowsill, or a commercial hydroponic farm, hydroponics allows you to eliminate a huge part of farming that you have little or no control over. Most hydroponic farms are indoor spaces and greenhouses where essential nutrients can be obtained and controlled. You do not have to worry about sunlight because there is artificial grow lighting that can serve the same purpose.

Chapter 2: Different Types of Hydroponics and Hydroponics Systems

Hydroponic system

Aeroponics

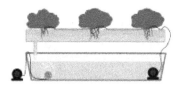

Nutrient film technique

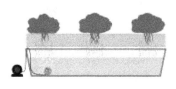

Wick system

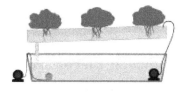

Drip system

There are various types of hydroponic systems. In this chapter, we will give an overview of the most commonly used ones.

The Drip System or Continuous Drip

One of the most popular growing systems, the drip system, is an active system that can be operated in either a recovery or non-recovery way. Pumps with supply lines that go to each plant are used in this growing system. A timer-controlled and standardized method of feeding the solution through drips is connected to the supply. The pumps are the mechanism that makes continuous drips possible. This method can easily be made into a recovery system by placing a tray under the plants to catch any unabsorbed solution and return it to the supply container for recirculation. However, recovery systems pose the possibility of causing the solution's effectiveness to lessen, as the reusing of it reduces its potency.

The benefit of the drip system is that it is easy to control the amount of moisture that is taken up by your plants. Consider using river rock as your growing medium if using the drip system.

Although they don't sound anything alike, the drip system shares almost the same equipment and design as with ebb and flow. This, of course, is made possible by attaching a drip line from the pump and then lining it across the medium. Gardeners can also add a drip manifold to have water drop from above.

Since the plants will not absorb every last drop of the water solution, the excesses will collect at the bottom. These need to be directed back to the reservoir, and that's why a drainpipe must still be attached.

It may sound silly and ineffective, but the drip system has its advantages over the ebb and flow. First off, it requires less water. Second, the water can be directed at the exact locations of the roots. Not only will this prevent the crops from drowning, but it will also save the gardener from frequent pH adjustments.

Recirculating systems are convenient in some ways, but they are the most prone to pH shifts. The higher the activity of the plant, the greater the fluctuations of the pH. Naturally, once the plant had fed on the water solution delivered by the pump, it will fall back to the reservoir with its pH already changed. And, once it gets mixed with the rest of the solution, the entire liquid's pH level will change.

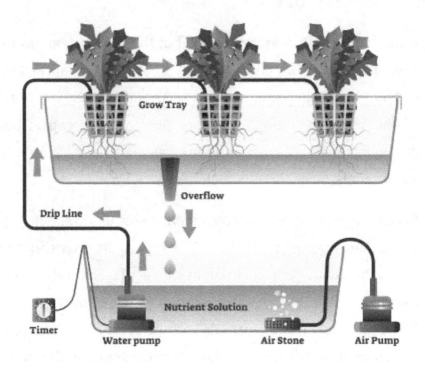

The Ebb and Flow System

The ebb-flow system, which involves a flood-and-drain concept, is another hydroponics system that is popular among many growers of hydroponics plants. It is easy to build, uses almost any accessible material, and will not cost you that much.

In the ebb-flow hydroponics system, your plant's root systems are flooded on purpose with the nutrient solution. This flooding is done on a periodic basis, not continuously.

To build the ebb-flow system, you will need these materials: A submersible pump (pond/fountain); tubing that will run from the reservoir (via the submersible pump) to the area to be flooded; an overflow tube set (make sure it is set to your preferred waterline), and a light timer for controlling the on/off function of the pump.

It is in the central part of your ebb-flow system that your hydroponic herb and vegetable containers are placed. A timer sets the submersible pump in motion by turning it on, and this action causes the nutrient solution or water to be pumped through tubing that goes from the reservoir to the ebb-flow system's main section.

The nutrient solution will not stop flooding the ebb-flow system until it becomes level with the overflow tube's preset height (about 2 inches below the growing medium's surface) and, as a result, causes your plants to get soaked.

There are two ways for how you may choose to plant your crops: the usual method–in separate baskets–or planting the crops all together in a tray.

The setup of the first option closely resembles the system used by the recirculating water method. The difference, however, is that the water here needs to be wholly drained right after the pumping stops.

The second option, on the other hand, will make hydroponics gardening a little easier. There will be no need to attach line after line of hose because the growing tray may sit directly above the reservoir. It will only require a tube leading to the tray, and another one where the water can drain back to the reservoir.

Since there is a wide space to fill, the best growing medium to use here are pebbles or gravel. This will be heavy, of course, so make sure to use sturdy materials. One advantage of not limiting the space of the plants is that you can opt for crops with wider root spans. Should you decide to go for such crops, make sure to allow a wide enough space for each.

Collect the remaining materials needed for the system after deciding on a planting method. Aside from the water reservoir, planting basket or tray, growing medium, air pump, and air stone, you will need a fountain pump and a timer.

With the use of the fountain pump, the goal of the system is to flood the plant baskets or tray with the solution water. Again, to keep it from overflowing and thus spilling, attach an overflow tube in the opposite end of the tray. How long the system should be flooding the tray or basket depends on the type of crops and the amount of water being pumped.

Gardeners should be careful about choosing the type of growing medium. Although pebbles and gravel are recommended, the more significant determining factor is the type of crop. If, however, the gardener does not have a wide choice, then what he would need to adjust is the frequency of watering. Pebbles and gravel keep the roots well aerated but they don't hold water. The only solution to prevent the crops from suffering is to have the pump work twice or thrice a day.

If you choose to grow your hydroponics herbs and vegetables in an ebb-flow system, keep in mind that you need to avoid using a growing medium that tends to float. Perlite and vermiculite are good examples of such growing mediums.

The Water Culture Growing System

Home gardeners and commercial, large-scale growers alike find the water culture system a truly effective way of growing plants hydroponically.

Besides being a concept that is not too difficult or too complicated to understand, the water culture system is also an extremely inexpensive hydroponics system to build. This is the reason many beginners in growing hydroponic plants at home see the water culture system as an ideal means to try their hand at soil-free gardening.

In using the water culture system, it is important to see that your growing medium does not take up too much moisture, to the point that it becomes saturated. What you are aiming for is to keep the bottom moist and the top dry – this will encourage your plants roots to grow downward and reach for the nutrient solution.

You can always use your imagination to build a water culture system using a number of different materials (you don't have to limit yourself to the ones presented below).

This is one of the simplest forms of hydroponic growing. It is a floating garden. However, instead of floating on ordinary water, the plants are placed on Styrofoam platforms and placed in a pool of nutrient solution. Pumps are still used, not to supply plants with solutions, but to provide oxygen. In a way, it is a non-recovery method because all the nutrient solution is placed in a large container so that the plants can float on it, with no recirculation being needed. After the plants have consumed almost all of the supply, the solution can be easily replaced.

What will complicate this a little, however, is the air pump that would need to run from under the reservoir. Aside from the water reservoir, plant basket, and growing medium, you would need an additional air pump, airstone, and air hose to set this up.

Again, you can opt to build this system from scrap materials. But since this will make use of electrically powered items, it's a wiser decision to buy newer or sturdier materials instead.

The ideal medium for this method are oasis cubes because before the crops can be transferred to this type of system, the plants first need to develop long enough roots. There will be nothing, after all, to deliver the water to the growing medium. And that means the roots themselves need to be submerged into the nutrient water to feed.

It turns out that roots are a lot like fish; they pick out the dissolved oxygen (DO) in the water to breathe. But just like the nutrients, the plants will exhaust the liquid's supply of DO. Thus, it needs to be constantly replenished; otherwise, the crops will drown. The good thing about DO is that it doesn't actually take much to produce it. Other than pumping air and creating minute bubbles, there are other much simpler ways of resupplying oxygen in H_2O.

Falling Water

Think of waterfalls in this method, and the churning it creates at the bottom. That's an example of nature supplying its bodies of waters with heavy loads of dissolved oxygen. Basically, any kind of disturbance made from the surface will add DO. Of course, however, lesser volumes of water and shallower agitations mean fewer DOs.

The problem with falling water is that it's better suited for large hydroponics systems. Just imagine how much effort it would require to design a series of waterfalls in your home. Besides, small crops wouldn't need that much oxygen to survive. If, however, you are planning on growing heavy feeders, like root crops, then the falling water method might work for you.

Recirculating Water

This water culture method is a combination of falling water and ebb and flow, and it's more complicated than everything previously discussed. To circulate the water, a fountain pump, overflow tubes, and another series of hose need to be added to the list of materials.

Start setting this up by attaching the air pump and fountain pump to the central reservoir -- a bucket or any container without a crop feeding on top of it. Then line the buckets with the crops before the reservoir, and connect them all with the hose. These tubes will lead the excess water back to the central reservoir and the entire process will repeat from the beginning.

Aside from the DO created by the air pump in the central reservoir, additional oxygen shall be delivered by the falling water action created by the fountain pump. Another advantage of the recirculating water method is the fact that the gardener only needs to check the water solution in the central reservoir. You wouldn't need to go over each bucket to check the pH and make adjustments.

The Wick System

Another type of passive non-recovery growing system, the wick system does not use pumps and moving water. A wick is attached to as many areas as needed, making the supply constant through capillary action. This is the process in which liquid travels or crawls through the medium (in this case, the wicks), towards the growing mediums. This style is not advisable for many plants because large plants that absorb large amounts of solution can use up all of the solutions that the wick provides. Because of this, smaller plants are not able to absorb the nutrients that they need.

The wick system of growing hydroponics plants does not rely on any air pumps or any special components to bring water to your plants' roots; rather, it depends on the moisture it is able to wick or take up through the fabric.

As a beginner in hydroponic gardening, the wick system may be the ideal system to work with, since there are no moving parts, pumps, or electricity for you to have to deal with.

When using the wick system in growing your hydroponics herbs and vegetables, you still have the option to use an air pump in your system's reservoir.

The wick system is a useful hydroponics system to work with for people living in areas where electricity is unreliable.

If you have kids who are interested in learning about hydroponics, you will find that the wick system is an excellent vehicle for explaining to them the process of plant growth.

Building your own hydroponics garden using the wick system only requires you to gather the following materials: A bucket to serve as the reservoir, another bucket to serve as your plant's container, a piece of wicking rope or several strips of felt fabric, and a growing media (perlite, vermiculite, and coco noir are good choices).

Working with the wick system is relatively easy. As its name implies, and with the help of some capillary action, you allow the nutrient solution that is in the reservoir to be wicked up to your plants. To put it simply, the wick system brings water to your hydroponics herbs and vegetables by sucking up water like a sponge.

This growing method is the simplest among the others. Therefore, if you are a hesitant beginner, wanting only to explore hydroponics, then this is the best place to start. The only materials needed for the wick system are the following:

Water container or reservoir

Growing medium

Wicking material

Plant basket

There is almost no expense in setting up the wick system. Scrap home items can substitute for everything listed above. The water container and plant basket, for example, can be replaced by an empty water or soda bottle. Simply cut it in half, use the bottom as the reservoir, then turn the top upside down, and use it as the basket.

You still have the option of buying brand new items for each listed material. Plastic bottles, after all, allow for only a single head of lettuce, or stem of herbs. If you want one reservoir to support several slots, then go for wide plastic storage containers. Drill six round holes on its lid, measuring at least 2 inches in diameter, then buy plant baskets in hydroponic stores. These baskets are about the size of yogurt cups, black in color, and slatted at the bottom and sides.

You can also use your imagination to design a different way of how the system will be put together. Just remember, however, that the system's primary function must still work in your design.

If, however, investing time and effort in creating your own from scratch is impossible, then opt for ready-made systems instead. Just like the plant basket, however, you may have to specifically visit hydroponics stores since not all gardening or horticulture shops have opened their shelves to this product line yet.

The logic behind this is simple. Keep in mind that the most essential function of a growing method is its delivery of the nutrient solution to the plant. As its name suggests, the wick system wicks the water up to the growing medium. This means two things:

The wicking material must be highly effective at channeling the water upward. You can measure how much it can transport by dipping a strip of the material in a glass of water, then connecting it to a smaller container above. You will eventually have to know this because different crops need different amounts of water. By seeing how much one can wick, you'll be able to estimate how many strips or strings will be needed to supply the right amount to the growing medium.

The perfect growing medium for this system will depend on the wicking material and the crop. If the plant requires large amounts of water, or if the wicking material is not as efficient, then a medium with excellent absorption and retention qualities should be selected. On the other hand, if the water requirements of the crop aren't that great, then a medium with balanced absorption and aeration qualities will be enough.

Chapter 3: Plant Nutrition

As you decide whether to build or purchase a hydroponic home unit, doing a little nutrient analysis would be a good idea. Regardless of what type of system you select, nutrients will be an integral part of your success because your plants need to be supplied with food continuously.

You can mix your own nutrients in large or small amounts using the formulae in this chapter. I suggest, however, that the novice starts with a commercially available, premixed nutrient at least until a hydroponic sensation has been established.

In soil farming, nature does a lot of the work, although often not completely, otherwise, farmers would not have to use fertilizers. Nearly all soil has nutrients in it, but you take over from nature when you grow hydroponically, and often you can improve the quality of the nutrients supplied.

Homemade Nutrients

The most common homemade nutrient is made from salts extracted from fertilizer. Such salts are available in bulk from farm companies, suppliers of plant food, some nurseries and gardening shops, and suppliers of chemicals. The only problem with this approach is that you generally have to buy these salts in twenty-five to fifty-pound bags, and unless you grow in large hydroponic gardens, these amounts will make the whole thing very difficult and expensive.

The following information should provide good general knowledge of these materials for the adventurous, or for the person who simply wants to know.

The salts marked with an asterisk are the best to work with where other, identical salts are available as they have superior properties such as enhanced solubility, quality, and a lifetime of storage and stability. For example, potassium chloride can be used instead of potassium sulphate, but if added for more than two days, the chlorine in the mix may be harmful to your plants. This is true because the chlorine level in your water is likely to be there in the first place. Magnesium nitrate can be substituted with magnesium sulphate, but using a costlier substitute for cheap and readily available Epsom salts does not seem worthwhile. Compared to cold water for ferrous sulphate, ferric citrate must be dissolved in hot water.

The secondary nutrients are: sulfur, iron, manganese, zinc, copper, boron, magnesium, calcium, molybdenum, and chlorine. The following list provides each individual nutrient's specific function in plant growth. In order to grow and develop, plants need nutrients, water, sunlight, and of course air. Although no soil is used, plants still need an anchoring medium that will sustain the weight of the plants through the roots. This is provided by a growing medium, which has the role of only offering physical support. Another function fulfilled is to block sunlight from reaching the roots, which will inhibit their growth.

In a hydroponics system, the water and nutrients are provided as a mix. This can be tweaked so that it fulfills the specific needs of a particular plant species. Light can be provided naturally or through special growing lights.

There is a multitude of different growing mediums as well as numerous mixes–some commercial, some created by the horticulturists themselves. However, all of them have to fulfill several basic requirements.

No nutrients or chemicals – the growing medium should not affect the plant in any way or interfere with the provided nutrient mix. It should be light and soilless.

Porosity – the medium should contain "holes" or "interstitial spaces" that will facilitate the easy transfer of oxygen and nutrients to the roots. The size of these empty spaces will vary with the size of the growing medium's particles. Thus, coarse growing mediums such as gravel have large spaces whereas sand has small spaces.

Control for pH – although not a requirement, a good growing medium will help buffer pH changes over time.

Safe and reusable – a growing medium should be reusable and biodegradable in order to ensure safe disposal. It should also be lightweight and easy to work with.

Price – another important factor that is not completely a requirement is the price. A growing medium should be inexpensive and easy to obtain.

Below you will find a selection of the most commonly used growing mediums, together with their advantages and disadvantages.

Perlite is a synthetic growing medium that has been used for many years. It is made from heated silica (glass) which expands into small and light pellets.

Advantages

It has excellent moisture and oxygen retention, as well as being chemically neutral. If added to a mix, it will increase moisture-holding capacity without adding weight. It can be cleaned with bleach and used many times over.

Disadvantages

Due to its lightweight nature, it can be easily washed away so it is impossible to use in ebb and flow systems. This can be avoided if used as a mix blended with another medium.

Vermiculite is similar to perlite in its properties. This growing medium is mined in Brazil, China, Zimbabwe, and South Africa. A naturally occurring crystalline compound, it is expanded by heating in a kiln. During this process, it becomes very light and water retentive.

Advantages

Like perlite, vermiculite is a good moisture retainer and chemically inert, thus has a neutral pH. It is lightweight and relatively inexpensive. It does not break down and can be reused if cleaned correctly. If mixed in a 1:1 ratio with perlite, it provides a lightweight, low cost, chemically neutral growing medium suitable for most projects.

Disadvantages

When used by itself, it retains too much water and thus can suffocate the roots of plants.

Coconut coir is another popular growing medium that is a natural byproduct of the coconut industry. This medium is the hairy outer coating that surrounds the coconut shell. It has the role to protect the seed from the sun and salt while floating around the ocean. Sold in blocks, it will swell to between six and eight times its compressed size.

Advantages

It has excellent moisture retention, up to eight times its own weight. It has the water retention of vermiculite and the air retention of perlite while being completely organic.

Disadvantages

Similar to perlite, due to its lightweight nature, it can be easily washed away. Thus, it is not suitable for an ebb and flow system. As with all growing mediums, it is rinsed in diluted bleach, rinsed again, and dried. This process can be done only three to four times before the coir begins to break.

LECA: or light expanded clay aggregate is a growing medium obtained from clay particles. These are lightly heated until they expand to somewhere between six to eighteen millimeters in diameter.

Advantages

It is a lightweight, free-draining medium that is still fairly good at retaining moisture. It is pH neutral and reusable. Due to its porosity, it provides very good oxygenation.

Disadvantages

Due to its porosity, this medium is inferior to perlite or coir in its water retention properties. Thus, it requires to be frequently re-watered or to be used in a mix. It is also expensive. Although reusable, the cleaning and sterilization process is complex and requires several steps.

Nutrients

Nutrient solutions are the cornerstone of every hydroponic system. The health of a plant depends on several components that can be visualized in the form of a chain. Oxygen, carbon dioxide, water, light, macronutrients, and micronutrients all represent a vital link in this chain. Thus, it is vital to ensure that all components are in place and in good supply. The nutritional needs of a plant are complex and creating a balanced solution may seem a daunting task. However, this becomes easier with some basic information.

Vital elements

There are four chemical elements that are vital for plants: oxygen, hydrogen, nitrogen, and carbon. These are also mostly overlooked when it comes to planting nutrition as they occur naturally. However, depletion of any of these would cause the death of a plant.

(O) Oxygen provides the energy that plants utilize to grow.

(H) Hydrogen comes from the water, and with sunlight as an energy source, it plays a vital role. It is also important in plant-soil relations. Furthermore, it is also essential for the formation of sugars. Water also keeps the plant's structure

rigid through what is known as turgor pressure. When a plant is lacking water, it will begin to lose turgor pressure and wilt.

(N) Nitrogen is obtained from the soil through the roots as amino acids or different types of ions.

(C) Carbon is obtained during photosynthesis from carbon dioxide. It is the element found in the largest quantities and represents approximately half of a plant's dry weight.

Plant nutrients

Macronutrients - Named so because they are required by plants in large amounts. They play a vital role in the growth and development of plants as they assist in different metabolic processes. Oxygen, nitrogen, hydrogen, carbon all belong here and are four of the nine macronutrients needed by a plant to survive.

The other five macronutrients are:

(P) Phosphorus is used to synthesize nucleic acids (DNA, RNA, ATP, etc), phospholipids, and sugar. It enables food energy to be converted into chemical energy.

(K) Potassium helps to regulate stomatal opening and closing, which maintains a healthy water balance. It is also involved in protein synthesis that needs potassium.

(S) Sulfur is part of the amino acids such as cysteine and methionine.

(Ca) Calcium regulates nutrient transport and supports enzyme functions. It is also required for cell wall formation.

(Mg) Magnesium is used in the photosynthetic process, chlorophyll production, and enzyme manufacturing.

Nitrogen, phosphate, and potassium are the components always listed on nutrient packets or bottles. Their proportional quantity is always provided always in the same order N-P-K.

Micronutrients - Although required by the plant in small amounts, they are still essential to its survival and are involved in critical functions.

Cations are positively charged and bind to soil particles. Their solubility is greatest under acidic conditions.

(Cu) Copper activates enzymes and is necessary for photosynthesis as well as respiration.

(Fe) Iron is involved in chlorophyll formation as well as the respiration of sugars to provide growth.

(Mn) Manganese is a catalyst in the growth process and formation of oxygen in photosynthesis.

(Zn) Zinc is involved in chlorophyll formation, respiration, and nitrogen metabolism.

Anions are negatively charged and subject to leaching.

(B) Boron is necessary for the formation of cell walls when combined with calcium.

(Cl) Chlorine is important for photosynthesis where it is part of the opening and closing of stomata. It also helps ensure that the leaves are firm.

(Mo) Molybdenum is part of nitrogen metabolism and fixation.

USE OF NUTRIENT SOLUTIONS

It is actually quite simple to use nutrient solutions. Only when you build your own nutrient blends and fine-tune the nutrition of your plants will it get complicated.

There are dozens of brands of different solutions for use, all with their own instructions, ratios, and programs. Some foster flowering, while others are mixed to foster increased root growth. Based on the brand's instructions, solutions are mixed with water to the right amounts and then used in your hydroponics system. Most nutrients are one-part, so there is only one substance to be mixed with water. More in-depth products are two or even three-part products that allow for more tailoring.

You will also find some fairly basic formulas bought as a two-part solution due to potential chemical reactions between ingredients, not because you can (or should) tailor your final solution by adjusting each portion of your use. Some compounds will react with each other when concentrated, causing crystals to form and emerge from the solution. You can prevent these types of reactions by first diluting each component and then mixing them. Be sure to read the instructions on how to do this.

Once a solution is formulated and used to fill the tank of your system, more nutrients must be applied as time goes by and the plants eat from the liquid. This can be tricky because you can't easily tell how many compounds of nutrients remain in the water after it has been circulating for several days.

Chapter 4: Actual Plant Growth

To all flower lovers who are planning to grow plants without soil, Hans von Berlepsch's appeal to "first of all study the theory" in full so as not to remain handicraftsmen for the whole life fully applies. This is true: everyone can purchase a special hydro pot, plant a beautiful plant in it, and take care of it in accordance with the instructions. However, in this case, there is no understanding of the relationships and hidden processes. In order to know the life processes of a plant well, this is clearly not enough, and it is such knowledge that is of the greatest value to us.

How Hydroponic Gardening Works - Growing Plants In and Without Soil

Soil has been closely associated with agricultural production since time immemorial. In the broadest circles, it has been taken for granted even today that humus-containing natural soil, with its infinite variety of tiny organisms, is an essential condition for normal plant growth. We affirm that we can grow plants just fine without soil, and we will try to substantiate this statement.

We consider the soil, that is, the upper layer populated by plants, the weathered surface of the globe, as a system characterized by three phases: solid, liquid, and gaseous. Any soil can serve as a habitat and a source of nutrition for plants when there is a favorable combination of these three phases.

In ripe soil, the ratio of these quantities, i.e., solid, liquid, and gas, corresponds to a proportion of 50:25:25. Half of the soil volume thus consists of a porous space, which again is half-filled with soil solution and half with soil air.

Solid soil constituents are predominantly solid inorganic materials. They are a product of weathered rocks ranging in size from large fragments to the smallest particles. The organic part of the solid phase of the soil consists of decomposed products of animals and plant organisms and the metabolic products of animals and microorganisms.

Natural soil is characterized by an endless variety of microorganisms that feed on the organic part of the soil. During this process, organic matter is completely decomposed to form water and carbon dioxide, and the mineral food products of plants contained in the organic mass are converted into a form in which they can be absorbed by plants. Along the way, microorganisms, due to complex chemical and biological processes, contribute to the further weathering of inorganic particles, and new quantities of plant nutrients are released. Thus, we can say that the totality of organisms living in the soil fulfills an extremely important task. In combination with other factors (various weathering factors), they continuously replenish the sources of nutrients in the soil. During the so-called mineralization process just described, plant nutrients such as nitric, phosphoric, and sulfuric acids are formed, which form salts with calcium, potassium, magnesium, etc.

The formation or release of vital trace elements (boron, copper, manganese, etc.) occurs in exactly the same way. All of these chemical compounds are important for plant nutrition but can only be absorbed by them with water, which serves as a means of dissolution and movement. Thus, soil moisture is a nutrient solution containing substances that are essential for plant nutrition. It must again be emphasized that the source of plant nutrition is only the soil solution with the nutrients it contains. On the contrary, organic compounds can be considered as sources of nutrients only after their complete microbiological decomposition. (Organic matter, of which the dry matter of plants is about 95%, is formed by the plant itself from water and carbon dioxide with the help of solar energy. They are never extracted from the soil in the finished form. The soil only supplies the missing 5% of mineral compounds.)

It should not be forgotten that water is necessary, not only as a solvent and a vehicle, but also as a nutrient in the construction of plants. In addition, water performs other various psychophysiological tasks (i.e., it promotes the swelling of colloids, etc.). No plant is able to grow without water, and in general, life without water is impossible. Lack of soil moisture can greatly reduce yield. Now, about the soil air. It should play a rather large role, because we always strive to promote aeration by cultivating the soil. This is understandable given that every living thing breathes, and therefore, requires oxygen. This, of course, applies not only to plant roots and storage organs (tubers, bulbs, etc.) but also to other organisms in the soil. If the surface of the soil coalesces so that normal air exchange is hampered, or if excess water in the soil displaces the soil air, then the underground parts of the plants suffer from a lack of oxygen. In this case, animal organisms inhabiting the soil can compete with cultivated plants with respect to oxygen consumption. Therefore, we must always take care of it.

We discussed very briefly how ripe soil should look, how plants would develop in the best way, and what fertile soil should be. From the foregoing, we can deduce the conditions necessary for also growing full-fledged plants without soil.

First, each plant requires a habitat in which it can be fixed by roots. It does not matter whether the roots will be settled in the mass of rice husks, gravel, peat chips, or coal slag. The substrate performs only a physical role and has nothing to do with plant nutrition. For this, a so-called nutrient solution is used.

A nutrient solution, as a natural source of plant nutrition, should contain all the compounds that a plant needs for lush growth and fruiting in the right form, sufficient concentration, and in proper proportions. Countless experiments with nutrient solutions have made it possible to clarify the needs of well-known cultivated plants so well that we can now make recipes for nutrient solutions. Periodic continuation of applying the solution and regular monitoring and replenishment of the loss of individual components allows us to provide good nutrition to our plants.

Microorganisms inhabiting natural soil are completely redundant when growing plants without soil due to the use of a ready-made nutrient solution. From it, plants receive all the food they need in an already digestible form; there is no need for the nutrient processing step. The nature of an artificial substrate does not need any influence from soil microorganisms. (In natural soil, we are very grateful to the organisms living in the soil for the formation of so-called soil aggregates.) Thus, we can choose materials that, after appropriate preliminary processing, will correspond in their structure to the structure of ripe soil (50% solid particles, 50% porous space). By this, we provide a fairly good supply of oxygen to the root growth zone, and thanks to the method of supplying the nutrient solution (we will learn more about this below), we can achieve an optimal supply of air.

Summarizing the above, we state that plants can be grown without any soil. You only need to be able to observe and simulate the processes occurring in the soil. If we can provide our plants with everything that is found in fertile soil, then we will achieve the same goal - the lush growth of healthy plants.

Why Plant Growth May Stop

If this happens, then you should immediately remember the "law of the minimum." What is meant by this?

Imagine the walk of a family with both younger and older children. The family moves quite slowly forward, because the short legs of the youngest children determine the pace of movement for the group. We are able to formulate a law: the speed of the family is limited by the feet of the youngest child - it is a limiting factor!

With the development of plants, similar circumstances play a role. The development of a plant is determined not by the growth factors that are available in optimal amounts, but by those that are lacking, which, therefore, are at a minimum. For this reason, even the best fertilizers and irrigation will not provide good results if you are trying to grow a light-loving plant in the dark.

The fact is that there is not enough growth factor that determines the boundaries of a plant's development, even when there are optimal quantities of other factors. This is called the "Law of the Minimum."

One clever and humorous gardener taught his students to always remember the five letters if they want their plants to grow. He had in mind the capital letters of the names of factors that are crucial for plant growth, LWAHN: light, water, air, heat, and nutrition. If a plant is provided with all these factors, it can fully manifest itself; that is, its growth will be most magnificent.

Using the method of growing plants without soil, we can directly affect the supply of plants with water, nutrition and, with a known skill, bring it closer to optimal. However, we should never forget about the other factors - light, warm air, and, as far as possible, we will take into account the special needs of individual decorative and healthy plants. These factors should not be limiting. There is a lot of good literature available for a more detailed look at these issues.

Chapter 5: Outdoor vs. Indoor

Hydroponics: Comparing Indoor and Outdoor Growing

Hydroponics is an interesting way to grow plants and crops. It is, of course, one of today's most groundbreaking farming innovations. Some of the benefits include water efficiency, better control of diet, high volumes of yield per any given volume of space, ease of automation, and more. You can grow crops outdoors or indoors using hydroponics technology. This chapter discusses how the two different approaches to growth vary.

Lighting

Installing artificial lighting will require a bit of installation and monthly maintenance money. The only benefit to artificial lighting is that the illumination can be controlled as much as you want. Otherwise, natural lighting is less expensive and requires no maintenance.

Temperature Regulation

Temperature is a critical development condition. Outdoors, temperatures are determined by the season. When you want to plant all year round, you can mount artificial heaters for the cold seasons at a discount. Indoors, automatic temperature regulation is absolutely necessary around the clock; it's one of the highest costs you'll have to account for. The upside again, just like with lighting, is that you can control the temperature to ensure maximum yield.

Space Economy

When it comes to agriculture, the air is important. Whether growing indoors or outdoors, you can take advantage of vertical space to create mediums of multi-level production. Cost and construction specifications in both sites (indoors and outdoors) will be almost identical. When it comes to space economy, the only difference is that there is more outdoor space available for agriculture than indoor space, particularly hydroponics.

Cost of Space

Another expense you need to calculate is that of space rented. Indoor space is generally more expensive because you are using a developed space that isn't easy to get through. Outside, in non-urban locations, you can find far less expensive land.

Safety & Durability

Indoor hydroponics has better odds when it comes to security and resilience than outdoor hydroponics. This is because the growth facility is covered indoors with mortar, concrete, or bricks. An indoor facility has much better chances of not being ruined in the event of a storm or vandalism attempt. In fact, indoor facilities are likely to survive better in the long term, while outdoor facilities may need to be refurbished every now and then, particularly the greenhouse coverings.

First, seek advice from a consultant on the hydroponics. They can give you a more specialized analysis, depending on where you plan to grow and what you plan to grow.

Types of Indoor Hydroponics

The Spare Closet Garden

When you live in an area where outdoor space is difficult to get through, here are some suggestions for building a "closet garden." This 6-inch PVC device suits perfectly within the space available. You can note that this setup fits well because of the two-level arrangement to maintain a constant supply of herbs, salad greens, and flowers. On the lower level, a 40-Watt fluorescent light is used to launch seedlings and root cuttings that are held flat at the bottom left within the 10"x20" humidity domed. If you want to take cuttings for faster growth and stronger stock, you can use the remaining area to grow a "baby" plant that is used for cuttings only. Once the seedlings or cuttings are rooted well, you can easily transplant them to the top of your closet for placement in your modified PVC system and high-intensity discharge (HID) lamp exposure.

The Do-It-Yourself Greenhouse

Consider building your own greenhouse if you're handy with a hammer and saw. By doing an internet search for "greenhouse ideas," you will find several plans and design ideas. It's a job that will take two people a weekend or so to finish, but will give you 10 or 20 years of service when using quality materials. The zoning board in your area may need to review the construction of any type of structure on your land, especially if you are building on a concrete slab. If clarification or change is needed, the city normally should ask the nearest neighbors if they disagree. For this intention, before breaking ground, I visited my neighbors to fill them in on my plans. I wound up with several more mouths to feed, so it goes without saying. They never knew I was going to feed them anyway. It's a hydroponic garden, after all!

The Professional Greenhouse

Over the years, I have worked with several people who had the privilege of being able to afford a homebuilder to have a professionally designed and developed greenhouse built. What makes these glass houses "professional" standard is: (a) they are made of structural steel and/or aluminum and glass instead of wood; and (b) insulated glass is sometimes used, if necessary.

What to Plant Outdoors

Tomatoes are the top choice for hydroponics outdoors. We regain the taste long gone from supermarket products, and the average yield in principle is very high, about twenty pounds per seed. However, in your first season, if you get around half that much, you're doing well. A medium-sized, fast-growing staking variety grows more and better quality fruit than the larger varieties. In hydroponics, tomatoes are grown fairly close together, about four to six inches. A reasonable design in a 16" x 24" unit would be to place a line of five to six plants along the rear edge and fan them out on lines or trellises against a wall.

Tomatoes need to be pruned on staking. It makes it easier to bind the vines together and to hold the plants in a sustainable shape. There are three pruning methods: single-stem, double-stem, or multi-stem. The best method in my experience is double- or multi-stem pruning.

It defined intercropping and outcropping, and the outdoor season is the time to take full advantage of your hydroponic growing area. The rest of the room is still usable for a vast array of flowers and herbs. Be sure to pay stringent attention to the rules provided in Chapter 7 on Companion Planting. It is important to get a good head start on the incredibly short growing season in cooler, more Western areas.

Another successful configuration is about nine tomato vines around the circumference of a 16" x 24" tank in your outdoor area.

When growing your hydroponics garden outdoors, make sure you keep your hydroponic device on a table or stand. If you don't, creepy crawls of all kinds will invade your plants.

It's time to move your hydroponic garden back indoors towards the end of the season. Because each latitude and location is unique, find out when the first frost is predicted from your local agriculture department. You'll still have an extra month to grow after your soil gardening friends at the end of the season, because of your building's security and heat and because your planters aren't sitting out on the table.

Cut the root stock from the growing medium before bringing your garden indoors (and whenever you pick an entire plant). Also, carefully check your plants and planters to make sure that no insect infestations are taken indoors. Otherwise, there are no specific directions to take back into your yard. The plants you want to keep will survive quite well, and the danger of temperature or light shock is small.

Keynotes for Outdoor Hydroponics

- Gradually move your garden from outside.

- Consult your newspaper for frost notices early in the season, and either cover your plants or move them for the night indoors.
- Begin your seedlings indoors two months before you begin your next step under the lights.
- Protect the seedlings from direct sunlight for the first few days.
- Prevent heat loss and provide safety by putting your plants against a wall. The maximum view is to the south or west.
- Work tall plants and mount them on poles, chains, or trellises.
- Keep your air pumps out of the rain by using an air hose longer than you would indoors.
- Search for rodents before relocating indoors.

Chapter 6: Inexpensive Hydroponics

Hydroponics is a simple method for growing plants without soil. It saves garden space, is less expensive, and is very beneficial in many ways. The hydroponics method increases fruit yield, requires less space, and is both profitable and ecological. Indoor gardening is possible with the help of hydroponics. You can have total control over the temperature, humidity, plant nutrition, and the overall environment of your hydroponic garden. If you have even a basic knowledge of gardening, you can successfully build your own hydroponic garden! Wheatgrass is a nutritious plant that can offer you many health benefits. You no longer have to go to the market or farms and waste your time collecting good quality wheatgrass. With hydroponics, you can plant your own healthy wheat garden. Hydroponics is the cleanest method of producing wheatgrass. With the right nutrients and water supply, you can successfully grow wheat in your hydroponic garden. You will be surprised by the rapid cultivation of wheatgrass using hydroponics! You can grow wheatgrass indoors with hydroponics at any time of the year and as often as you want. To build a simple hydroponic garden with wheatgrass, all you have to do is follow these basic steps for hydroponics gardening:

- As with traditional gardening, you should start hydroponic wheatgrass gardening by buying seeds from a local dealer.

- Place the seeds evenly in the container so that they cover the entire base of the container.

- Immerse the seeds in water halfway. Note, do not completely cover the container with water.

- Now you can cover the drawer with a second drawer or another plastic cover because the seeds don't need light at first, but make sure you don't close the drawer because the seeds do need air.

- Water your hydroponic wheatgrass daily with a spray bottle and as soon as you see that your plants have started sprouting, add water slowly as your plants grow. In addition to this simple, homemade cultivation method for wheatgrass, you can purchase a hydroponic grow set from your local hydroponic store. A wheat-growing set contains all the components you need to grow your hydroponic wheat effectively. In very little time, with a small initial investment, you can have your tray full with fresh and healthy wheatgrass. Be careful when choosing your hydroponic wheat kit because not all trays guarantee the freshness of good quality hydroponic wheatgrass!

Hydroponic Herb Garden

The nutritional requirements of hydroponic herbs can vary from plant to plant. It is possible to grow several hydroponic herbs in a single nutrient solution, but special attention must be paid to the prevention of hydroponic herbs against nutrient deficiencies. There are three main types of hydroponic herbs for different purposes; for example, there are herbs that can be used specifically for cooking (culinary herbs), there are medicinal herbs (home healing), and ornamental herbs. They can be grown in a single unit for a year without setting up the unit to grow again, as opposed to growing hydroponic fruit and vegetables where you need to clean the hydroponic unit after each harvest to start over. You can build your hydroponic garden specifically for cooking. Hydroponic herbal systems are automated and easy to use. The three main types of simple hydroponic growth systems are the food film technique (NFT), the ebb and flow averages, and the aerologic agents. It can easily be done anywhere, indoors or out, with a simple gardening method. For successful gardening, you have to understand the desired pH of the plants you grow, the soil needs, water needs, possible pests, etc. Herbs grown in hydroponics appear to be healthier, tastier, and fresher than traditionally grown herbs. These herbs are free from chemicals and fertilizers. Medicinal hydroponic herbs are simple and reliable. Hydroponic marijuana is an important ingredient in therapies to treat AIDS and cancer; it also works as a painkiller under certain circumstances. Hydroponics herbs have been used for decades in therapeutic programs. Another medicinal benefit of marijuana is in the treatment of glaucoma, a condition that causes vision loss. Growing hydroponic marijuana is a wise choice because it helps to cure many diseases while sitting at home. Medicinal gardening can easily be done at home and can be used immediately in tea and extracts.

Hydroponic grow sets for the indoor gardener

Many people turn to the garden to reduce food costs for their families and to make contact with nature. However, if you live in an area where outdoor gardening is out of the question, such as an apartment building or a hostile environment that is too dry or too cold for plants to thrive, it may be a bitter disappointment to want to do gardening, but without a real way to do it. This is when hydroponic culture sets can come in handy. Hydroponics packages help the novice gardener succeed with indoor hydroponic gardening, as they all contain the materials you need to start this exciting hobby. Although hydroponics packages vary from manufacturer to manufacturer and depending on the type of hydroponics gardening you want to do, some basic components are common. Because every type of plant needs water, food, and light to grow and thrive, you want to ensure that hydroponic systems reflect these needs. There are hydroponics lamps available in kit form, which is an easy way to start building hydroponics systems. Hydroponics lamps can be LEDs or HID lamps, which require a digital ballast to function properly, as well as reflectors and possibly fans, depending on the number of lamps installed and the size of the room. Because hydroponics is a form of aquatic gardening where no soil is used, each of the hydroponics packages you are considering must have a way to retain water or place water around the roots of the plants. However, the crown of the plants must be kept out of the water so that they do not rot, and this is achieved with hydroponic supplies such as nets or trays with holes in it so that only the roots are submerged in the water. Pumps and filters are used to move the water and help supply oxygen. Pumps can also be used to raise and lower the water level depending on the system you choose to use. Lights and hardware are typical components that are included in all hydroponic kits. A final item that you will need to ensure proper growth and production in your plants is food, which is called hydroponic nutrients in gardening. Hydroponic nutrients contain all the macro and micronutrients that your plants need. There are hundreds of formulations from which you can choose the formula that best meets the needs of the specific plants that you want to grow.

Why hydroponics?

Hydroponic gardens are compact and can be placed anywhere.

- They use and reuse water repeatedly and need a limited amount of extra water to function properly.

- They eliminate the need to control garden pests such as aphids, caterpillars, potato beetles, and fungi.

- They are very efficient growers for plants - plants grow very quickly in a hydroponics configuration.

- They are practical and most systems are easy to automate, so they require minimal interference from you. Every plant can grow (or start growing) in a hydroponic system, regardless of the time of year or which hemisphere you are located.

There are other reasons why hydroponics is preferable to a traditional garden, but there are also a few disadvantages. For example, many people associate hydroponic gardening with the growth of certain illegal plants that are often used as controlled substances. It seems that every week, the police are destroying another large house in a nice neighborhood. These houses are full of hundreds of compact fluorescent lamps, water sprays, containers, dust, nutrients, and plants that are removed. However, like everything, a small percentage of people can ruin something good for everyone. In reality, the main benefits of hydroponics gardening are to give people who would otherwise be unable to grow plants the opportunity to grow plants. It is very common for passionate gardeners to start their delicate young plants in a hydroponics configuration and then transfer these plants to their regular gardens after the soil has thawed.

Orchid growers, in particular, seem to be inclined towards hydroponic cultivation systems. The obsession that many people have with orchids is intense. This obsession, combined with the frustration of not being able to meet the demanding needs of the orchid in the unchanged backyard of a person, encourages many to try to grow in greenhouses or hydroponic facilities. Moreover, hydroponics technology is everywhere. Light timers are used in many applications to save energy, just as they are used in hydroponics to time the light cycle of plants. Compact fluorescent lamps, metal halide, T5, and other types of intense lighting used in hydroponics are also used in aquarium systems that strive to meet the demanding needs of freshwater plants or delicate corals and anemones. Water drop systems are regularly used in greenhouses and large-scale agriculture, but also in gardening and landscaping.

Plant nutrients have been developing for quite some time - as long as people are trying to grow non-native plants in partially impoverished soils. The pH meters are being used in scientific applications as well as in all forms of gardening. We would not know where the acidic soils are that would grow the grapes if we did not use a pH tester. Hydroponics has grown out of a combination of need and desire. We want fresh tomatoes, we want fresh basil, and we need a place to grow them because we don't all live on farms. The more we hear about plants covered with wax and pesticides, the more we worry about these substances that end up on our tables and in the food supply meant for our children and ourselves; we want a way to be sure that our food sources are safe. Although it is impossible to think that we can go from buying our food in the store to growing everything in our apartments, it is good to know that we can supplement some of our products in this way. Special sauces, for example, made from your herb garden of fresh coriander, basil, and oregano are a tempting reason to switch to hydroponics. For pure and hard natural gardens, hydroponics may not be your cup of tea. If you grow native plants from your region, simply sow your seeds and let nature take its course. However, for many people in northern climates, waiting for spring is too far away and the growing season is much too short. These people appreciate the plant life that technology can bring into their homes. Something to keep in mind: Before I knew what hydroponics was, I bought "tomatoes aged on vines and grown with hydroponics" from the local supermarket because of their color, texture, and superior taste. They were constantly bright red, juicy, and free of stains. They were always smaller than the so-called "stew pot tomatoes," but the larger, lighter-colored tomatoes had a less rich taste. The balance of nutrients, light, and water, combined with the reduced need of the plant to fight infections, insects, and fungi, produces a much healthier plant specimen than would otherwise be possible.

Forget Organic - go hydroponics!

People will pay more for organic products. But did you know that there is an even cleaner way to grow all your fruit and vegetables at home, without paying a penny in the supermarket? Market leader, Great Stuff Hydroponics, wants to explain how this is possible. Known as hydroponics, the science of growing plants without using soil is becoming the next big thing among environmentally friendly consumers with green fingers. This method can be used to grow plants all over the world in every season. As a result, plants are naturally healthier than their soil-grown counterparts (geoponics). Many commercially grown hydroponic plants do not require pesticides and are certified organic. Hydroponics of plants also has many environmental benefits. The hydroponics method uses only about 1/20th of the water used for irrigation on traditional farms to produce the same amount of food.

Normally the "drain" of farms ends up on the water surface. Hydroponic growth therefore considerably reduces the effects of large-scale agriculture on the water level, with the added advantage that the water used and its effects on the surrounding land can be accurately measured. Hydroponic plants are already commercially grown on a large scale in Israel, Nicaragua, and the United States, where people can grow fresh produce with limited water supply in arid landscapes. Going to buy or grow hydroponics fruits and vegetables also has certain benefits for the consumer; there is a shorter delay between picking and packing the plant, which means that it stays fresh longer. Moreover, because hydroponic plants have better access to the sun and better nutrition, the products are healthier and tastier. Steven Parker, director of Great Stuff Hydroponics, says, "Growing hydroponic plants is something anyone can do at home with one of the Great Stuff Hydroponic kits, for beginners or advanced growers. Because of the methods used, you can get all your favorite vegetables and fruit, out of season at home."

How to mix hydroponic nutrients

Pour just enough water into a tank, tub, bucket, or another container to fill the tank with nutrients for your hydroponic system. Be sure to measure the number of gallons.

Check the pH value of the water that you are going to use in your nutrient mix, using a pH test kit for hydroponics systems.

You can buy this kit from hydroponics stores and suppliers. The pH must be between 5.0 and 6.0. If the level is outside this range, use the chemicals in your kits to adjust the pH level up or down until it is within this range.

Measure two teaspoons for every gallon of a complete dry water-soluble fertilizer. Use a fertilizer with micronutrients such as magnesium, sulfur, potassium, nitrogen, phosphorus, and calcium. Also, make sure it contains essential micronutrients such as boron, zinc, chlorine, cobalt, manganese, iron, copper, and molybdenum. Then add this fertilizer to the water.

Add one teaspoon of Epsom salt for every liter of water.

Mix all the dry ingredients in water until all the crystals and powder dissolve in the well.

Remove and replace your nutrient solution every one to two weeks to ensure that the plants continuously have the nutrients they need.

Advice

If your tank contains a lot of water in your hydroponics system, mix one or two extra batches of hydroponics. Never overfill the container, otherwise, the hydroponic nutrients will spill by mixing. Use the solution immediately after mixing, as it may lose some strength if you store it for use.

Warnings

If the water level falls between changes due to absorption or evaporation, only use normal plain water to replace what has been lost. Adding a new batch of nutrient solutions to the old one can cause an imbalance and salts that will damage the plants can accumulate. This type of fertilizer is best used in ebb and flow hydroponics systems, hydroponics raft configurations, and other hydroponics systems that do not tend to clog. Hydroponic systems such as geoponics or drip hydroponics can become clogged when solids reach the nozzles.

Hydroponic Garden Backyard vs. Hydroponic Greenhouse

The growth of plants in the courtyard of your hydroponics may not seem ideal in the use of a hydroponic greenhouse. A hydroponic greenhouse offers better lighting and irrigation on the configuration of the system. Not many people will support the greenhouse concept if they are used to a hydroponic garden system. You need sufficient space for the installation of irrigation systems and lighting systems that are needed for hydroponic gardens. Where is space available in this fast-paced world full of houses and skyscrapers? But if you have a hydroponic greenhouse, you can place these systems much easier and also in a smaller space. Lighting and other arrangements are supplied with the greenhouse, so you don't have to worry about those factors. Most plants that do well under hydroponic conditions are carefully examined.

Chapter 7: Which Specific Materials/Equipment Do You Need?

PH Meter

The pH level is used as a measure of how acidic or how alkaline water is. A pH of 7 is neutral. PH levels that range from 1 to 6 are acidic, and levels from 8 to 14 are considered alkaline or basic.

Different plants have their preferences regarding optimal pH levels. To ensure the best possible growth, you need to have a way of testing and then adjusting the pH level of your water.

For example:

- Cabbage likes pH levels of 7.5
- Tomatoes like a pH of 6-6.5
- Sweet potatoes like a pH of 5.2-6
- Peppers like a pH of 5.5-7
- Lettuce and broccoli like a pH of 6-7

We will talk about why balancing pH is essential later in the book.

A pH meter can be obtained from local hydroponics stores or online. You need to calibrate the sensor with the calibration powder that comes with the meter. A basic pH meter will cost you $10 to $20.

A basic pH meter

Don't use paper test strips for the water because they are inaccurate. Most of the time, a pH meter is offered in combination with a TDS or EC meter, which we will talk about next.

EC Meter

Electrical conductivity is a measurement of how easily electricity passes through water; the higher the ion content, the better it is at conducting electricity.

All water has ions in it. When you add nutrients to the water, you are increasing the ion content, effectively increasing the electrical conductivity.

EC or electrical conductivity is an integral part of the hydroponics equation. The simplest way of explaining this is as a guide to measure salts dissolved in water. Its unit is siemens per meter, but in hydroponics, we use millisiemens per meter.

In short, the higher the number of salts in the water, the higher the conductivity. Water that has no salt (distilled water) will have zero conductivity.

Lettuce likes an EC of 1.2 (or 1.2 millisiemens), while basil likes an EC of 2.

A $15 TDS & EC meter is available from Amazon.

That is why it is so important to know your EC and what your plants prefer; it will help you to ensure your system is at the right level.

Additionally, electrical conductivity needs are also affected by the weather. When it is hot, the plants will evaporate more water. For this reason, you need to decrease the EC in hot summer months.

- In warm weather, you need to decrease the EC.
- In cold weather, you need to increase the EC.

An EC meter won't tell you the specific amount of which mineral or fertilizer is in the water. If you only use a nutrient solution using the right ratios, you shouldn't worry.

Just because it doesn't monitor individual nutrients, doesn't mean that it's not useful. Salt levels that are too high will damage your plants, regardless of which salts they are.

You generally need to keep them between 0.8 and 1.2 for leafy greens and between 2 and 3.5 for fruiting crops like tomatoes. The source of the water can also influence the EC reading. More on this later.

Sometimes, you see the recommended nutrient levels listed as CF. CF is the conductivity factor. This is like EC, and also primarily used in Europe. If you multiply EC by ten, you will get CF.

For example, lettuce grows best in an EC of 0.8 to 1.2. This is equivalent to a CF of 8 to 12.

TDS Meter

TDS stands for total dissolved salts. You may hear some hydroponics growers referring to the TDS rather than the EC. These are both used to determine the strength of your hydroponic solution. If you buy a TDS meter, there will also be an option to switch to EC readings.

It is crucial to understand that TDS is a calculated figure. TDS readings are converted from EC readings. The problem occurs when you don't know which calculation method was used to produce the TDS; there are several different ones.

In general, EC and CF readings are primarily used in Europe, while TDS is an American measurement. But, regardless of which measurement you choose to use, they are both effectively the same thing: a measure of the nutrient levels in your solution.

The NaCl Conversion factor

This is effectively measuring the salt content in the water. The conversion factor for this mineral is your microsiemens figure multiplied by any number between 0.47 and 0.5. You'll find most TDS meters use 0.5. This is the easiest one for you to remember and calculate. Most of the meters sold will use the NaCl conversion factor.

As an example, if you have a reading of 1 EC (1 millisiemens or 1000 microsiemens), you will have a TDS reading of 500 ppm.

Natural Water Conversion Factor

This conversion factor is referred to as the 4-4-2; this quantifies its contents. It is forty percent sodium sulfate, forty percent sodium bicarbonate, and twenty percent sodium chloride.

Again, the conversion factor is a range, this time between 0.65 and 0.85. Most TDS meters will use 0.7.

For example, 1 EC (1000 microsiemens) will be 700 ppm with a TDS meter that uses the natural water conversion.

Potassium Chloride, KCl Conversion Factor

This conversion factor is not a range this time. It is simply a figure of 0.55. Your EC meter reading 1EC or 1000 microsiemens will equate to 550 ppm.

These are not all the possible conversion options, but they are the most common. The first, NaCl, is the most used today.

Dissolved Oxygen Sensor

Plant roots need oxygen to remain healthy and ensure the plant grows properly. The dissolved oxygen sensor will help you to understand how much oxygen is available in the water and ensure it's enough to keep your plants healthy.

If plants don't get enough oxygen to their roots, they can die. A minimum of 5 ppm is recommended.

A dissolved oxygen (DO) meter will be expensive for the hobbyist to buy, especially when you are starting. That is why people who do hydroponics for fun and hobby purposes generally do not purchase DO meters. A good meter can cost you $170 to $500 for a reputable brand.

You do not need to invest in one if you oxygenate the water. Oxygenation of the water can be done by using an air pump with an air stone in the water tank. Depending on the method of growing, you don't need to aerate the water.

The dissolved oxygen in the water will be at its lowest during the summer. The water heats up, and the dissolved oxygen becomes less available. While your plants can do very well in winter, they might lack oxygen during summer.

Net Pots

In some systems, you are going to need net pots to hold the plants. This is mostly true for deep water culture (DWC), Kratky, wick systems, Aeroponics, fogponics, Dutch buckets, and possibly vertical towers.

Make sure you get the net pots with a blip on top to keep them from falling through. The standard size for lettuce is 2 inches (5 cm). If you want to use tomatoes with Dutch buckets, 6 inches (15 cm) is recommended.

3- and 2-inch (7 and 5 cm) net pots

If you are creating a new system on a budget, there are a variety of other options that can be used instead of buying net pots. For example, plastic cups with lots of holes in them, or simply fine netting on a wireframe. Use your imagination!

6-inch (15 cm) net pots for a 5-gallon (18 liters) bucket

Humidity and temperature sensor

Estimating temperatures and humidity levels will lead to mistakes. I recommend getting a simple humidity and temperature sensor, so you don't need to guess. Most of them will cost you no more than $15.

Germination Tray and Dome

You need to start seeds in a dedicated germination tray. Most of these trays are 10" x 10" or 10" x 20" (25 x 25 or 25 x 50 cm) and generally include a humidity dome.

These trays are used to let your seeds germinate and keep the humidity high. After the first true leaves appear, it is time to transplant them into your system. Usually, this is after ten to fifteen days.

The humidity should be between sixty and seventy percent, while temperatures should be 68-77°F or 20-25°C.

Seed Starter cubes

If you are growing plants from seed, you can't simply place seeds in the net pots. They'll get washed away or sink. Instead, you need a seed starter cube. These cubes provide a place for the seed to start growing roots and flourish, safely.

Several materials can be used for media when starting seeds:

Rockwool Cubes

These are made from a combination of basalt and chalk, spun together. The result is a small cube that is similar in consistency to cotton candy. Your seeds can be placed into the Rockwool cube where they will start to germinate. The cube goes in the net pot and your seed should have everything it needs to start growing, providing it has access to the nutrient-rich water. These cubes come in all shapes and sizes.

Generally, you would only need a small 1" – 1-1/2" cube. Depending on the method of growing, you need to add some grow media to support the cube and block sunlight from creating algae on the cube.

You can separate each cube from the bigger sheet. I recommend using gloves for this because the Rockwool can be irritating on the skin.

As with every seed starter cube, you must soak it in pH neutral water for six hours before using it. This ensures that the seed has better germination rates.

Coco Coir

An alternative to Rockwool cubes is coco coir. This is simply the fibrous coat of a coconut.

Coco coir is an organic media that will break down over time. Some people use it because it is environmentally friendly and renewable. I don't recommend it to start with. It can break down and clog your system if you are not careful. It can begin to rot, and before you know it, make your water quality terrible.

Oasis Cubes

Another seed-starting cube is the oasis cube. They retain moisture well and are very soft. This makes it easy for the roots to penetrate the medium but also makes them brittle. Oasis cubes are also used as a growing media. These cubes are very popular with NFT systems.

Sponges

Sponges are used most of the time as a cheap alternative to Rockwool or oasis cubes. However, they do not absorb or retain moisture that well. That is why using sponges is not a carefree method of seed starting. They are not as environmentally friendly as the other seed starting cubes.

Growing media

After you have placed your seed in the seed starter cubes and they have started to germinate, you will see roots coming out of the starter cube.

This is the time for you to start transplanting them into their second grow space. The grow media will depend on which growing technique you are going to use.

For floating rafts and NFT, you do not need any growing media. For other methods like the Kratky, Dutch buckets, or wick system, you may need to add some growing media.

The use of grow media also depends on how big the plant is going to be. Lettuce doesn't need growing media because it doesn't need the support. Tomatoes need growing media to support the plant.

Growing media gives your plant stability and space for the roots to further develop.

Hydration

This is the most popular growing media.

Hydration is created from clay that has been heated to high temperatures. The result is a very porous material that is made into small balls.

Hydration is very lightweight, ensuring your pots aren't under any undue stress. It is excellent for keeping your seed starter cubes in place. It's also easy on the hands.

You should wash hydration before using it to remove any clay dust.

You can also re-sterilize this grow media, but this can be time-consuming, especially if you have a lot of it.

Sterilizing or re-sterilizing is vital as your grow media can have bacteria or other micro-organisms that could be harmful to your plants. Whether using it for the first time or re-using the growing media, cleaning alone may not be enough to get rid of these bacteria. You need to sterilize the growing media to ensure it is safe for re-use.

Sterilizing involves using either heat or a chemical to kill all organisms on the growing media. A popular chemical choice is hydrogen peroxide.

You will need a thirty-five percent hydrogen peroxide solution and then wash the clay balls in it thoroughly. Mix one part of hydrogen peroxide (thirty-five percent) with eleven parts of water. This will lower it to a three percent solution.

Most importantly, you need to rinse the growing media several times to ensure all traces of the hydrogen peroxide are gone.

You could also use a ten percent bleach solution. A bleach solution is used for sterilizing NFT troughs or other equipment in your system as well.

The alternative is to heat the growing media in an oven with a temperature of 180°F (82°C) or more, for at least thirty minutes. It will pasteurize and remove fungal type microorganisms. To get rid of all organisms, you need to sterilize them, which involves at least thirty minutes at 212°F (100°C).

Warning: Doing this in your kitchen will leave a sour odor that lingers in your home.

Perlite

When you take minerals, such as found in volcanic glass, and expose them to extreme heat (1600°F or 871°C), you will force them to expand and pop like popcorn, effectively creating Perlite. This is another growing media that is pH neutral and extremely light. It's not good at retaining moisture.

Perlite is used by gardeners in their soil to increase aeration to the roots. It is excellent for Dutch buckets (drip), wicking systems, or the Kratky method. More on these later.

You can get Perlite at any garden store, but you must be careful not to get it in your eyes.

When handling perlite, use a dust mask. The dust created when handling perlite is not healthy to breathe in.

Vermiculite

This silicon-based substance is exposed to the same high temperatures for forming Perlite. It also expands and is very similar to Perlite, and it is also pH neutral. The main difference is that Vermiculite is high in cation-exchange.

In simple terms, it is better at holding onto water and nutrition for release into the plant later. Vermiculite is so good at retaining moisture that it can suffocate the roots. That's why it's most popular as a 50/50 mix with perlite.

Rockwool

As already described, Rockwool is an excellent choice for seed starting. The larger Rockwool cubes that range from 3 - 6 inches (7 – 15 cm) can be used for an entire plant. It is pH neutral and easy to use. Rockwool is very good at wicking up the water.

The big Rockwool cubes are mostly used for plants that have a long-life cycle. It is not financially viable to plant lettuce in these bigger grow blocks. The holes in these blocks are too big for seedlings. You need to use a Rockwool starter cube to grow your seedlings, and then move the plant with the starter cube into these bigger cubes.

Another method of growing is using Rockwool slabs. They are widely used in tomato farms. They are the largest Rockwool media you can find. You can fit several big Rockwool cubes in one slab. One slab is a few feet long and comes in different sizes. You need to pre-soak these too.

Do not remove the plastic cover from these. The plastic keeps the moisture in and the light out. Drainage holes should be made at the bottom of the slab. Drip emitters should be inserted into the big Rockwool cubes.

Grow Stones

Grow stones are made from recycled glass, which may seem like an unusual material for a hydroponics system.

Grow stones are good for aeration and moisture retention. The fact that they can wick water up to 4 inches (10 cm) above the waterline means your plants will always have the water they need.

You shouldn't let the term glass put you off; they look like sharp edges, but they are not.

River Rock

As the name suggests, this type of grow material comes from a riverbed. The rocks are naturally rounded off as the water removes the sharp edges. The irregular shape of river rock means that there are plenty of air pockets, which makes it easier for the roots to become established.

River rock must be washed before you use it to ensure it is clean. If you are thinking of using this, it is worth noting that it is heavy and cumbersome, potentially preventing you from moving your system in the future.

Pine Shavings

This is not the same as sawdust, which will absorb water and block up your system. Pine shavings must be made from kiln-dried wood, and there must be no chemical fungicides in it. Your best bet for purchasing this inexpensive grow media is pet stores or a nearby wood processing factory.

Chapter 8: How to Build Your Hydroponic System

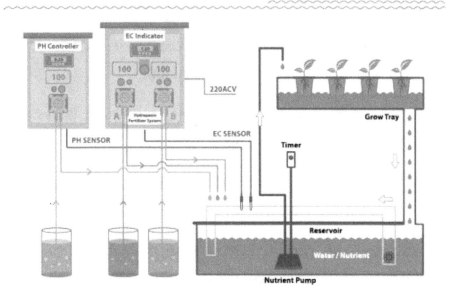

Steps Involved

First, line your Styrofoam box with a black bag and then fill it with distilled water.

With your cups, trace six evenly spaced circles on the lid cover (the deep part). Using a knife, cut out circles that are a ¼-inch smaller than the traces.

Cut a small hole in the bottom of each of the Styrofoam cups and fill it with moss and perlite (just to under the cup's rim).

Sow lettuce seedlings in the perlite and fill it in.

Water the seedlings every day with the nutrient mix. You should do this at the same time each day so you will not forget. Always check that the water level in the container is not too high.

After 7 to 10 days, you will see roots spring out from the seedlings and they will enter the container.

As the lettuce matures, harvest it as you need it. Remember not to over-harvest or the plant will die.

When you are successful with this first hydroponic garden, you can move on to try other plants and other systems.

Setting up the hydro unit is also easy and you can purchase the nutrient mix from gardening stores. One of the tricky parts is knowing how to germinate the hydroponic seeds and transplant them in the hydro units so your hydroponic garden is ready.

Tools and Everything You Need

As you venture into hydroponics farming, there are specific tools you will need to get started. Although there are several systems to choose from, the kind of tools used in all of them is more or less the same.

The tools you will need include:

Reservoir

From its name, the reservoir is used for reserving the nutrient concentrate. The concentrate is typically a mixture of water and the required plant nutrients. Depending on the kind of hydroponic system you choose to install, the liquid is pumped periodically from the reservoir into the growing chamber as set by the timers.

In some systems, the reservoir doubles as the growing chamber too, such that the plants grow by suspending their roots in the nutrients concentrate at all times.

You do not have to purchase a special reservoir; you can fashion it from almost any inert large container that you use to hold water, so long as it does not leak. The container should be able to hold enough of the solution to allow the plants to grow. In addition, the container should be opaque to prevent the rays of the sun from penetrating into the solution.

If the container available to you is not opaque, there are many ways to make it light proof. For example, you could wrap or cover it up with an opaque material, or you could paint over it. The idea behind the opaqueness is to prevent algae from growing on the inside of the container.

If the hustle of making your own reservoir seems a bit too much, you could also opt to purchase a commercial reservoir and they will serve you well.

Growing Chamber

A growing chamber is one of the most critical parts of a hydroponic system because this is where the plant roots develop. The chamber is the container that holds the roots, provides support to the entire plant, and houses the nutrient concentrate.

The chamber, just like the reservoir, should be kept from direct sunlight and extreme temperatures because these can introduce heat stress to the plants. In cases of exposure to extreme temperatures, such as heat, the plants abort their fruits and flowers.

The size and shape of the growing chamber is dependent on the kind of hydroponic system you intend to run and the plants you wish to grow. Plants that grow big roots will require a large growing chamber while those that develop small roots will be okay with just smaller one. However, do not be stressed about sizes because any chamber size will make due so long as the plants you are growing get their deserved nutrients and space.

In your quest to find the best growing chamber, avoid metallic containers because metals are subject to corrosion and they will react with the nutrient concentrate. If you cannot purchase a commercial growing chamber, however, check around to see what non-metallic items you could transform into growing chambers. If you need to maintain class and style while at it, you could opt for a commercial growing chamber; there are some fabulous models available, and I am certain they will appeal to you.

Delivery System

The delivery system is the system that delivers nutrients to the plant roots directly. The concept of this is quite simple, in fact, and it can be customized to fit into any system you choose to install. A typical delivery system must include connectors, PVC tubes, blue or black vinyl tubing, and tubing connectors for garden irrigation.

Depending on the hydroponic system that you settle for, you can choose to use emitters and sprayers for the delivery system. Although sprayers and emitters are quite useful, be prepared, however, for frequent clogs when the nutrients in the solution build up in the system. Therefore, if you are looking for stress-free farming options, avoid these as best you can.

Submersible Pump

Most pumping systems have a submersible pump to regulate the pumping of the nutrient concentrate from the reservoir to the growing chamber. You can buy these pumps at home improvement stores or hydroponic shops in your area. The pumps come in varying sizes, and you just have to choose one that matches the size of your farm.

How do submersible pumps work, you ask? Well, the pumps are simply impellers that take advantage of electromagnetic fields to spin and then pump water. It is easy to maintain them because the majority of the time, you are only required to clean the solution filter. If you bought your submersible pump without a filter, you could still make one by cutting part of a furnace filter, ensuring that it fits the submersible pump.

Besides the filter, you also need to clean the pump occasionally to ensure that there are no clogs that would obstruct the nutrients as they flow to the plants.

Air Pump

Although it is not compulsory that you make an air pump part of your hydroponic system, you ought to give it a thought because it comes with so many benefits. Air pumps are widely available in stores, and inexpensive, particularly if you are able to buy yours at a store that sells aquarium supplies.

An air pump is primarily used to ensure that there is a steady supply of oxygen in the water so that the roots can absorb it for their respiration, in the growing chamber. The pump does this by pumping the air through air lines, onto the air stones, which then creates air bubbles that bubble up into the nutrient solution.

In case you are using a water culture hydroponic system, for example, the air pump keeps the roots from drowning in the nutrient solution because they are kept suspended in it all day, every day. In other hydroponic systems, the air pumps are fitted into the reservoirs to keep pumping oxygen into the water, increasing the oxygen concentration in the water.

Since the air pumps pump all day, they cause constant movement, which keeps the water and nutrients in it in constant motion. The circulation that results from the process ensures that the nutrients dissolve into the water evenly, at all times. The presence of oxygen in the water is also good because it prevents the growth of pathogens and microbes.

Timer

Not all hydroponics farmers need to time their operations with a timer. The decision to use a timer is based on the choice of hydroponics system and its location. If your system is to be situated indoors, for example, and you have installed artificial lighting, you need to install a timer that will turn the lights off and on.

Drip and aeroponics systems also need a timer to control the submersible pump that controls the process of draining and flooding. It is important to take note of the fact that some types of aeroponics would need a special kind of timer to work properly.

Although light timers and standard pump timers work very well, it is better to opt for a timer that has a 15 amperes rating rather than 10 amperes rating. A 15 amperes timer is heavy duty and will have a cover that effectively protects it from water. Check the back of the packaging to ensure that you have made a good choice.

For those who may have a battery backup, a digital timer is not preferred over an analog one because once you unplug it from the power source, it loses all the data previously stored in it. Analog timers are a better choice for the additional benefit of having on and off settings. Therefore, as you go out to purchase a timer, ensure that yours has pins all around the dial, so that you get the analog kind, and avoid future regrets.

Growing Medium

The growing medium is essentially the substance on which the plants grow. It provides physical support to the plants, just like soil does, except that it is inert, not containing any minerals or living organisms. Different systems demand different growing mediums. For example, while other systems use peat moss, Rockwool, or lava stone as the growing medium, an aeroponics system uses air as the growing medium.

Nevertheless, the right kind of medium is one that retains moisture in such a way that the water solution will not need to be pumped in continually, every single minute.

Growing Lights

When it comes to grow lights, you can have them or you can go without them. They are an optional part of the system because it all depends on where you intend to plant your garden. You may end up using natural light or having to take up artificial lighting for your plants. If possible, opt for natural lighting because it is free, and will not add to the cost of setup as you purchase the new equipment and its accompanying maintenance costs.

If, however, you cannot get enough good lighting in the place you intend to plant your garden, by having lots of exposure through the window or having a sunroom, or that the time of the year does not allow enough lighting through, you may need to include some supplemental artificial lighting in your setup budget.

Ordinary light bulbs cannot be used as grow lights. Grow lights are specially made light bulbs that emit light containing special color spectrums that mimic natural light. Your plants will take in these color spectrums and use them to carry on the process of photosynthesis, hence leaf growth, flower formation, and fruit growth. Realize also that the intensity and type of light that the plant has access to, by large, determines its photosynthetic abilities.

Most hydroponic kit systems will come with complimentary light fixtures, but if you are setting up a DIY (Do It Yourself) garden, piecing together the equipment you need, you will need to purchase your own lighting fixtures.

The most effective lighting for a hydroponics system is the high-intensity discharge (HID) light fixture made up of either metal halide (MH) bulbs or high-pressure sodium (HPS) bulbs. The HPS, in particular, emits a red or orange-looking light, which works well for plants, particularly in their vegetative growth stage.

Another type of lighting used is T5. It produces fluorescent light of a high output. This lighting consumes low energy and emits only a small amount of heat. The T5 is suitable for growing plant cuttings, and for growing plants with short growth cycles.

Ensure that the light is kept on a timer so that the lighting will go on and off at the same time, each day.

pH Testing Kit

If you don't test the pH of your nutrient solution from time to time, you will be running your farm purely by guesswork, subjecting your entire investment to a trial-and-error game. The reality is that for your plants to thrive in the hydroponic garden you have set up, there needs to be a balanced pH, and using a pH testing kit, you can regularly check on your garden to determine whether the pH of the nutrient solution is optimal. If the pH is too low, you can adjust by bringing it up, and if it's too high, you can lower it as well.

On a related note, besides the pH meter, you will also need equipment to measure the temperature and the ppm of the water. You can also purchase equipment to measure the humidity and temperature of the grow room. If, for example, you find that you need to adjust the humidity in the room, use a dehumidifier or a humidifier, to ensure that the plants do not dry out and that they do not stay too damp.

A fan or any other equipment that can be used to improve the air circulation in the room would also be welcome. Although a small oscillating fan may work for a beginner, you will need a more sophisticated fan as your garden grows, particularly one with an intake and an exhaust system.

The Nutrient Solution

While the nutrient solution is not technically a tool, you will need to set it aside as you set your tools aside, in readiness for the setup of your garden. As we have established many times so far, the nutrient solution will be the primary source of nutrients for your plants to thrive.

The nutrient solution provides three primary macronutrients that can be found in most fertilizers: potassium, phosphorus, and nitrogen, and a host of 10 other micronutrients that may not be found in the fertilizers; yet, the plants need them to survive, grow, and reproduce. Some of these micronutrients include zinc, molybdenum, boron, copper, iron, chloride, and manganese.

As a beginner, it may benefit you to purchase an already mixed solution offering a balance of all the nutrients mentioned above, but as you gain more experience, you will find it easier to create your own nutrient solution, one that will provide the plants with all the specific nutrients they require.

Chapter 9: Growing Mediums, Nutrients, and Lights

Light

In nature, plants are dependent on the sun's energy. Sunlight is converted into sugars through a process called photosynthesis, to provide food for the growth of the plant. These sugars are used in a process called respiration, and excess sugar is stored for later use. Chlorophyll, which is found within the leaf cells, makes photosynthesis possible. Chlorophyll contributes its signature green color to the plants. Light is trapped by chlorophyll, which triggers the photosynthesis cycle. Light energy is mixed with carbon dioxide and water for processing oxygen and sugar inside the chlorophyll. Through the respiration process, the sugar is then oxidized (or metabolized), producing carbon dioxide, water, and growth energy. The foliage transpires excess oxygen and water into the air. Therefore, plant growth is directly affected by the color, intensity, and duration of light obtained by the organism.

High-Intensity Discharge (HID) Lighting

Nothing beats the sun when it comes to production, but new types of high-intensity discharge lighting have made it a viable alternative for growing indoors. Many of you are familiar with the fluorescent "grow" lights designed to grow plants indoors. Such goods are ideal for low-light plants where tests are likely to be minimal. But what if you want your favorite indoor plants to achieve their ultimate growth potential? Or, you want to supplement your greenhouse with sunlight? High-intensity discharge illumination, or HID for short, is your best choice. Such lighting systems consist of a torch, reflector, and power supply and are designed to provide the maximum output for the amount of power absorbed by photosynthetically active radiation (PAR). With the right quality and quantity of light, HID lighting systems will illuminate your garden to achieve impressive results. Horticultural HID lighting is used by the world's leading farmers to provide traditional fluorescent and incandescent lamps with many advantages that are practically unattainable. HID illumination allows commercial farmers to raise crop yields, deliver crops to market on time, and harvest crops when they are out of season, making them even more attractive to customers. HID lighting is so effective and strong that even after the initial investment, many indoor growers make a healthy profit. Previously, horticultural HID illumination for daily gardeners was prohibitively expensive due to low demand and high production costs. Yet, thanks to suppliers such as Sunlight Supply and Hydro Farm's innovative new lighting devices, lighting prices have been reduced to the point where everyone can reap their advantages.

Intensity

The intensity of light is generally measured in power (watts) per square foot. A general rule of thumb for optimum photosynthesis is 20-50 watts per square foot, with 20 being better for low-light plants, and 50 best plants at least 24" unless the lamps are suspended by a radial or linear light mover, in which case you should minimize the lamp by 25-50 percent to plant size. Cover the growing area with a semi-flat white color, often referred to as an eggshell coating, to improve light effectiveness. The limited gloss in this form of paint will ensure maximum absorption whilst still helping you clean off smudges or stains that may occur over time.

Certain wall finishes include:

Mylar 90-95% reflective

Flat white paint 75-80% reflective

Gloss white paint 70-75% reflective

Yellow paint 65-70% reflective

Aluminum foil 60-65% reflective

Black < 10% reflective

Duration (Photoperiod)

Most plants grow fastest when exposed to 16-18 hours of light per day. It has not been observed that extra hours of light during the day improve development by any significant amount. Plants that exhibit photoperiodism should be exposed to 12-14 hours of light and darkness once flowering is desired, the trait that causes the amount of day-length to prompt flowering.

Light (Photosynthetic spectrum)

Photosynthesis is most pronounced in the wavelengths of light, red (600-680 nm) and blue (380-480 nm). Horticultural illumination, also known as high-intensity discharge (HID), is designed to cover these specific wavelengths, known as the PAR (photosynthetically active radiation) continuum. There are two types of HID lamps that emit different spectra of colors. Metal halide lamps emit a spectrum of white/blue. MH lamps are best used as the main source of light (if there is very little or no natural sunlight). This type of lamp encourages compact growth in vegetation. MH-to-HPS conversion bulbs are also available, allowing you to supply MH light during vegetative growth and then switch to HPS for growth stages of fruiting/flowering. Sodium lamps with high pressure emit a yellow/orange spectrum. These are the best secondary or supplementary illumination lamps available (used in combination with natural sunlight). This form of light encourages plant bloom/budding. HPS lamps are suitable for greenhouses and for increasing commercial applications. The Son Agro and Hortilux HPS lamps add an additional 30 percent blue factor to their spectrum, which makes them a better choice for solo use compared to straight HPS lamps.

Choosing a Grow Light

When choosing an HID lighting system, red and blue are the two main light colors that you'll need to consider. Blue light is most pronounced when the sun is highest in the sky during the spring and summer months. It is in charge of keeping plant growth small and shapely. Red light is responsible for causing the reproduction of plants in the form of flowers and fruits, for example, when the sun is lower in the sky during fall harvest months. Metal halide (MH) lamps mainly emit blue light, making them ideal for the stage of vegetative growth. High-pressure sodium (HPS) lamps mainly emit red light, inducing excessive flowering and fruiting during the reproductive period of the plants. So if you're planning on growing mostly leafy crops like lettuce and vegetative herbs, your best bet is an MH lighting system. Invest in a Son Agro or Hortilux HPS if you want to grow flowering plants as it adds about 30 percent more to the blue spectrum than a standard HPS.

Remember, lights emit heat to keep indoor gardens within the range of 65-80 degrees and 50-75% humidity. The primary benefit of using a horticultural lighting system for high-intensity discharge (HID) is the power that it provides you over the growing environment of your plants. This is a great benefit to those of us who like to have a year-round supply of fresh flowers, veggies, and herbs! HID illumination is also a great way to jump-start your spring seedlings months ahead of the last freeze. Another important feature of indoor horticultural lighting is your ability to control daylight lengths, thus empowering you with the ability to force your favorite strain to flower even when out of season altogether. Photoperiods for vegetative growth vary from 16 to 18 hours/day. Once you power the lights for 18 hours/day, the cost of electricity is minimally beneficial, and not worth it. Flowering photoperiods normally range from 10 to 14 hours per day. Remember, the secret to the perfect light for the right plants is Intensity, Duration, and Color!

Growing Plants

Now that we are here, we are sure that you are ready for this! It's time to use your new knowledge and begin to grow. We will discuss some of the most common cultivars in this segment and learn about their favorite growing environments. We will also learn about seed starts and how to take and root cuttings. Before we continue, however, let's quickly go over a few more essential plant needs.

Temperature

The rate at which plants grow is influenced by their ambient temperature. Normally, as the temperature rises, so do other facets of the plant's metabolism and, as described below, they may or may not be within the optimal ranges for either genetic factors or any other limiting factors. It is essential to keep your environment within the temperature range your crop needs to avoid stress and excessive ripening in order to achieve the best growth.

Humidity

Relative humidity is defined as the amount of water found in soil. High humidity levels can stop plants from transpiring water through their leaves, as the air is already full of water. High humidity can also keep plants from cooling down through the same transpiration cycle, and can harm them by providing the right atmosphere for growing powdery mildew and botrytis.

Light

Not all light is created equal, especially when it comes to plants. Light appearing within the color spectrum that activates photosynthesis is called PAR (Photo-synthetically

Active Radiation), and it's the only kind that will affect the crop's growth rate. Many light meters do not calculate PAR, which restricts the utility of the meter in calculating how quickly your plants are going to grow. Even if they don't weigh PAR, the majority of meters are useful in determining whether the light is even over your yard. The Moon, HPS, MH, and now compact fluorescent lamps all emit PAR illumination.

CO_2

Normal CO_2 concentration (325-425 ppm) can be rapidly depleted in enclosed environments, resulting in slow growth due to the lack of photosynthesis. Providing plenty of fresh air or extra CO_2 (ranging from 1000-1500 ppm) will sustain steady chlorophyll production, and plants will grow rapidly.

Dissolved Oxygen

Dissolved oxygen (DO) is the measure of oxygen available in your solution. Roots require oxygen to breathe and will fail if they are not consistently supplied with the correct amount of oxygen. Stagnant water must be disturbed or oxygenated in lakes and wetlands if plants are to be grown directly in them. A general rule of thumb is to preserve DO in solution, which directly feeds and bathes plant roots between 5 and 25 ppm. If the plant does not have the correct level of DO, anaerobic respiration will occur, which will allow the plants to generate toxic levels of ethanol rapidly.

pH

The pH of a solution is the relative number of hydronium ions contained therein. The pH level varies from 0 to 14, with a pH of 7 being neutral, 0 being highly acidic and 14 being extremely alkaline (or fundamental). If the pH is zero, the solution produces equal numbers of hydrogen ions (H+) and hydroxide ions (OH-) to even it out. A solution with a pH of 0 to 6.9 has a higher H+ ion concentration, which makes it acidic. In this case, however, a solution with a pH of 7.1 to 14 has a higher concentration of OH-ions, resulting in it being alkaline, or basic. The pH in any nutrient solution is highly critical, as it controls the availability of elementary salts. Nutrient deficiencies can occur at pH levels outside the standard due to their non-disponibility to the plant.

Total Dissolved Solids (TDS)

Electrical conductivity (EC) is explained as the measure of the total dissolved solids in a solution. Because nutrients are taken up by a plant, the EC ratio is reduced because the solution contains less salt. Conversely, as water evaporates from an open pool or is absorbed by the plants, the EC of a solution decreases.

Grower's Guide to Common Plants

Always use a high-quality hydroponic fertilizer and maintain a healthy growing area, allowing you to provide your plants with plenty of light, air, and moisture. Seed packets will provide more detail about the specific strain that you wish to grow.

Important Note:

The ratios of nutrient solutions shown here (PPM/EC) are only a general rule of thumb. When you work with a commercially available nutrient solution, make sure to follow the instructions of the supplier for use with the crops you choose to grow. This is primarily due to the variation between the materials used in the various industrial products.

Getting Started with Seeds

Most plants rely on the seed as their primary reproductive process. After pollination by the male flower, the seed is produced inside the female flower. All seeds emerge as an egg inside a female floral carpel. When wind or mosquito drops male pollen into the female flower, the egg becomes an embryo and develops a rough covering around itself. The seed is released when the seed production finally stops, and it is taken to its final resting place by wind, rain, moth, or insect. If all conditions are right, it will turn into a new plant and repeat the process of the growth cycle. I simply "tick" the open flowers with a soft artist's brush on the peppers and tomatoes to spread the pollen from one flower to other flowers.

I like using a growing area of 10" x 10" or 10" x 20" to give a friendly environment to the seeds and additional cuttings.

Keep the stickiness high by utilizing a 6" clear vault spread. A little creativity and some Tupperware and clear saran wrap will work as well.

You'll also need to choose a starting medium and a growing medium. The starting medium is the thing that you will plant your seeds or cuttings in until they develop sufficiently enough to be "transplanted" into the framework.

Typically, you will be starting your seedlings in a growing medium.

Perfect Starts wipes are produced using natural fertilizer that is formed into little fittings with an adaptable cover that keeps up there wipe like surface and shape. This permits the hydroponic cultivator to utilize a natural mechanism for seeds and cuttings that can be transplanted legitimately into a framework. The wipes secure the roots and shield the material from obstructing shower heads.

Abstain from utilizing soil to begin your seeds since it isn't sterile and may contain bugs or pathogens that can contaminate your framework. Water your starting medium with a 1/2 quality supplement arrangement before use and keep it soggy, but not drenched, while seeds or cuttings root.

If you are utilizing coco coir, it comes in dried out blocks that can be absorbed the 1/2 quality supplement arrangement during the re-hydration process.

See the section on growing modes for more data on re-hydrating. One block makes about two gallons of free coco coir, so you might not need to utilize the whole block immediately, unless you have a great deal of seeds to start.

Seed Starting

From my experience working with various methods for starting seeds, I have built up a basic and solid strategy for effective germination.

1. Pre-dampen the starting medium w/1/2 quality supplement, pH6.0.

2. Maintain a temperature of 72-80 degrees.

3. Keep the air temperature at 70-78 degrees and 70-90 percent humidity.

4. Use delicate light (20 watts/sq. ft.) until most grow, at that point increment.

5. Feed 1/2 quality supplement until light power is expanded.

6. Dispose of frail growing seedlings.

7. Move seedlings to the production area once leaves begin to sprout.

Chapter 10: Advantages and Disadvantages of Hydroponics

Hydroponic gardening is gardening at its best. There is little or no dirt involved in hydroponics growth. Hydroponic gardening is the use of water and light to grow fruit and vegetables. Hydroponic growth means less time spent and less money wasted on useless materials. You do not have to spend money on fertilizers and pesticides. You also do not spend hours on weeding and plowing. Hydroponic gardening is very beneficial because the yields on crops are much higher and plants generally produce richer, brighter, and more nutritious fruits. To start your hydroponic garden, you must decide where you will install your plants. The hydroponic growth of plants usually means that you need considerable space for plants to grow. Most people use a greenhouse. The hydroponic growth of plants is fairly simple and almost everyone can do it. You only have to do a little research, especially when you are just starting. Information is readily available. Talk to people you know who love hydroponics gardening. Discover what kind of nutrients your plants need. Hydroponic nutrients are generally more concentrated because they need to be added to plants and their growing environment. It is best to find a combined solution that contains all the nutrients needed to grow your plants. In this way, you can control not only the light and the water but also the number of parasites that influence the yield of the plants. When growing a garden outdoors, be prepared to lose some of your crops due to pests, weather, and other factors. However, with hydroponics, you can eliminate most of these factors. Different types of hydroponic nutrients can stimulate your plants to produce more flowers, which in turn yields more fruit from plants such as the tomato plant. It is just as necessary for plant growth as water and light. Discover the essential elements of hydroponics to guarantee a beautiful garden all year round. You will not regret it when you see the abundant harvest.

No Soil Required

With hydroponics, you can grow crops in places where the land is limited, non-existent, or heavily polluted.

Hydroponics has been considered by NASA as the agriculture of the future for growing food for astronauts in space (where there is no land).

Make Better Use of Space and Location

Because everything the plants need is supplied and maintained in a system, you can grow in your small apartment or in spare rooms as long as you have the space. Plant roots generally expand and spread in search of food and oxygen in the soil. This is not the case in hydroponics, where the roots are driven into a tank filled with oxygen-rich nutrient solution and in direct contact with vital minerals. This means that you can make your plants grow much denser, and therefore becomes a huge space-saving system.

Climate Control

Just as with greenhouses, hydroponics producers have full control over the climate - temperature, humidity, intensification of light, and air composition. In that sense, you can grow food all year round, regardless of the season. Farmers can produce food at the right time to maximize their commercial benefits.

Hydroponics Saves Water

Plants grown in hydroponics only use 10% water compared to those grown in the open field. With this method, the water is recirculated. The plants absorb the necessary water, while the drain-off is collected and returned to the system. Water loss occurs only in two forms - evaporation and system leaks (an effective hydroponics configuration will have minimal or won't have any leaks). It is estimated that traditional agriculture uses up to 80% of groundwater and surface water in the United States. Because water will become a critical problem in the future when food production is expected to increase by 70% according to the FAQ, hydroponics is considered a viable solution for large-scale food production.

Efficient Use of Nutrients

The nutrients are stored in the tank, so there is no loss or change of nutrients as happens in the ground.

pH Control of the Solution

You can measure and adjust the pH values of your water mixture much more easily than you can in soil. This guarantees the optimum growth rate. Do hydroponic plants grow faster than plants in the soil? Yes, they do.

No Weeds

Once you have grown anything in the ground, you understand how irritating weeds are for your garden. This is one of the most time-consuming tasks for gardeners - plucking, plowing, hoeing, etc. Weeds are primarily associated with the soil. Remove the dirt and your weed problems are gone.

Fewer pests and diseases

Just like weeds, getting rid of the soil helps make your plants less vulnerable to soil pests such as insects, birds, and groundhogs, as well as diseases such as Fusarium, Pithier, and Rhizoctonia. When cultivating indoors in a closed system, gardeners can easily take control of most of these threatening variables.

Reduced Need for Insecticides, Herbicides, and Other Chemicals

Since you do not use soil, and will not have weeds, pests and plant diseases are greatly reduced. This means that fewer chemicals are necessary. You will grow cleaner and healthier food. Controlling pH of the solution affects absorption of mineral nutrients for plants. You can measure and adjust the pH values of your water mixture much more easily than in soil. You can naturally increase quality and yield of your crop without using harsh chemicals to promote growth.

Work and Time Saving

In addition to spending less work on tillage, watering, growing, and killing weeds and vermin, you also save a lot of time because plant growth is increased with hydroponics. When agriculture needs to be based more on technology, hydroponics has a place in it.

Hydroponics is a hobby that relieves stress

Even without growing in soil, hydroponics brings you back in contact with nature. Tired after a long day of work and commuting, you return to the corner of your small apartment to relax and have fun with your hydroponic garden. Reasons such as lack of space are no longer sufficient. You can start fresh and tasty vegetables or essential herbs in your small cupboards and enjoy the relaxing time with your small green areas.

Cons and Challenges

A hydroponic garden requires time and commitment

Like all things that are worth doing, a diligent and responsible attitude produces satisfactory returns. In terrestrial counterparts, however, plants can be left alone for days and weeks and still survive. Mother Nature and soil will help regulate things if something is not in balance. This is not the case with hydroponics. Plants will die faster without proper care and knowledge. Remember that your plants depend on you for their survival. You must take good care of your plants and the whole system during the initial installation. To automate everything later, you must always assess and avoid unexpected operational problems and perform regular maintenance.

Experience and technical knowledge

You are using a system made of many types of equipment, for which specific expertise is required for some of the equipment used. You have to know about the plants that you can grow and how they will survive and thrive in a bottomless environment. Errors made in setting up the systems and growing capacity for plants in this groundless environment could ruin your progress.

Organic debates

There have been heated debates about whether or not hydroponics should be certified organic. People wonder if plants grown in hydroponics will get microbiomes as they do in the soil. People around the world have been growing hydroponic lettuce, tomatoes, strawberries, etc. for decades, especially in Australia, Tokyo, the Netherlands, and the United States. They have supplied millions of people with food. Certain organic cultivation methods have been proposed for hydroponics producers. For example, some producers supply microbiomes to plants using organic culture media such as coconut fiber and earthworms. Nutrients of natural origin are often used, such as fish, bones, alfalfa, cotton seeds, etc. For this debate on the issue of organic products, research is continually conducted and then we will know what works best.

Water and electricity risks

In a hydroponics system, you mainly use water and electricity. Be cautious of using electricity in a combination of nearby water. Always put safety first when working with water systems and electrical equipment, especially in commercial greenhouses.

The threat of system failure

Electricity is used to manage the entire system. Suppose you do not take any preparatory measures in the event of a power failure. The system stops working immediately, the plants dry out quickly, and they can die within a few hours. Therefore, a backup power source and a plan must always be in place, especially for large-scale systems where you have the most at risk.

Initial costs

You can spend less than a hundred to a few hundred dollars (depending on the scale of your garden) to purchase equipment for your first installation. Whatever systems you build, you will need containers, lights, a pump, a timer, culture media, and nutrients. Once the system is installed, the costs are reduced to only nutrients and electricity.

Long return per investment

If you follow the news about starting an agricultural business, you may know that new hydroponics companies have recently started. This is good for the agricultural sector and the development of hydroponics. However, commercial producers still face major challenges when they start large-scale hydroponics systems. This is important because of the high initial expenditure and the long uncertain return on investment. It is not easy to develop a clear and profitable plan to stimulate investment, while many other attractive high-tech fields seem promising enough for financing.

Pests and diseases can spread quickly

With hydroponics, you grow plants in a closed system with water. In the case of plant infections or pests, they can quickly degenerate into plants in the same food reservoir. In most cases, pests and diseases are not a problem in the small systems of home producers.

- Don't worry about these problems if you are a beginner.

- It can be a complicated problem for large hydroponic greenhouses.

- It's best to have a good disease management plan in place.

- Use only clean, disease-free water sources and growth materials; make periodic tests and verification of systems, etc.

- If diseases occur, you must quickly sterilize the infected water, replace nutrients, and clean the entire system.

Chapter 11: New Techniques in Hydroponics

Hydroponic techniques involve the harvesting of batches of small plants that mature early. They refer to the method in hydroponic gardening where smaller plants are grown over shorter periods instead of growing a few big plants over a long time. With hydroponics, everything in the environment is controlled, from lighting to ventilation. It is possible to start one batch at an earlier time, and as they mature, another batch is started. This method results in a year-round growing and harvesting cycle. Another way of doing this is by starting all the plants together and creating a green canopy where you let your plants be harvested more times than once. Taller plants will be harvested from the top first without uprooting them. As the plant grows more, the lower level becomes the top level that is ready for harvest.

Hydroponics and quality of vegetable products

Hydroponics production is the method for growing plants in soilless conditions with nutrients, water, and an inert medium (including gravel, sand, or prelate).

From a plant-science perspective, there is no difference between above-ground plants and plants grown in the soil, because in both systems nutrients must first be dissolved in water so the plants can absorb them. The differences lie mainly in the way nutrients are available for plants. In soil-based production, the elements adhere to soil particles and transfer to the soil solution, where they are absorbed by plant roots.

There are different types of hydroponics, depending on how they are characterized. A criterion is to classify closed or open hydroponics systems. Hydroponics systems that do not use culture media are generally referred to as closed systems, whereas hydroponics systems with culture media can be closed or opened in a container, depending on whether the nutrient solution is recycled (closed) or with each irrigation cycle (open).

Another approach to classifying hydroponics systems is to classify them according to the movement of the nutrient solution: active or passive. Active means that the nutrient solution is displaced, usually by a pump, and is passively dependent on a wick or anchor of the culture medium.

Others characterize hydroponics with criteria for recovery or non-recovery. Recovery occurs when the nutrient solution is re-introduced into the system, while non-recovery means that the nutrient solution is applied to the culture medium and then disappears. Although there is a wide range of criteria, there are three basic elements for plants: water/moisture, nutrients, and oxygen. All these different types of hydroponics must offer these three basic things that are important for the successful production of plants. Despite this diversity, the criteria most commonly used by producers, farmers, private companies, and researchers classify hydroponics systems in six different types: nutrient film technology (NFT), wicking system, ebb and flow (flood and drainage), aquaculture, the drip system, and the aerologic system.

Description of the hydroponic system

Geoponics this is the most advanced and high-tech method where plants are hung on special paths. The nutrients are sprayed directly onto the roots every few minutes, creating a light layer of nutrients. This system requires regular monitoring of the pumps to prevent malfunctions.

In a DWC (deep water culture) system, a tank is used to contain the nutrient solution. The roots of the plants are suspended from the nutrient solution to obtain a constant supply of water, oxygen, and nutrients. In this system, an air pump is used to supply the water with oxygen so that the roots do not drown.

In a drip system, the nutrient solution is set aside in a tank and the plants are grown separately in a bottomless environment. Drip systems release nutrients very slowly through nozzles, and additional solutions can be collected and recycled, or even disposed of. With this system, it is possible to grow several types of plants at the same time. The ebb and flow are also called "flooding and drainage." This is the least used system in hydroponics.

Like geoponics, the nutrient film technique (NFT) is the most popular hydroponics system. With this method, a nutrient solution is continuously pumped through the channels in which the plants are placed. When the power solutions reach the end of the channel, they return to the beginning of the system. This makes it a recirculation system, but unlike DWC, the roots of the plant are not completely submerged, which is the main reason for calling this method NFT.

The wick system is the simplest and easiest hydroponics method. It is a completely passive system, which means that nutrients are stored in a tank and introduced into the root system through capillary action. With this system, we can use different growing media such as prelate, vermiculite, coconut fiber, and other formulations. The wick system is easy and inexpensive to install and maintain. The main disadvantage of this system is poor oxygenation of the plant roots and a large amount of nutrient solution that is needed to effectively reach the plant root system.

Bioactive compounds

According to Biesalski et al., it is generally accepted that bioactive compounds are defined as essential and non-essential compounds. They are present in nature in the context of the food chain and have a positive effect on human health.

Bioactive compounds can be classified according to different criteria. The most commonly used classification in the literature is based on their pharmacological and toxicological effect. However, this is more relevant for clinicians, pharmacists, or toxicologists and not for plant biologists, agronomists, or other researchers involved in plant-related studies. For the latter groups, it is normal to classify them according to biochemical routes and chemical classes.

Example of health benefits of chemical classes

Glycosides (Cardiac glycosides, cacogenic glycosides, glucosinolates, and anthraquinone glycosides)

Recent reports provide a great deal of evidence about the chemo preventive role of different types of cancer glycosides (bladder, prostate, esophagus, and stomach) and degenerative cancer diseases. Glucosinolates, one of the most relevant groups of glycoside compounds, are capable of activating the enzymes involved in the detoxification of carcinogens, thus protecting against oxidative damage.

Phenol (Natural monophenols and polyphones, including phenol acids, flavonoids, aprons, chalconoids, flavonolignans, lingams, stableness, curcuminoids, and tannins)

Epidemiological studies have claimed that polyphones can capture reactive oxygen species (ROS), inhibit peroxidation lipids on cell membranes, prevent LDL cholesterol from oxidizing, and protect DNA from mutations and oxidations. These include fruit and vegetables, tea, cocoa, wine, grapes, and peanuts.

Arytenoids (Arytenoids with provitamin activity, which are vital components of the human diet)

Diets rich with arytenoids are connected to a significant reduction in the risk of certain cancers and cardiovascular and degenerative diseases. Epidemiological studies have shown that arytenoids can have anti-cancer and anti-mutagenic properties. These are found in green, orange, red, and yellow vegetables.

Plant sterols (Phytosterols)

They regulate the fluidity and permeability of phospholipid double layers in plant membranes. Plant sterols are believed to have anti-cancer, anti-atherosclerotic, anti-inflammatory, and antioxidant activities.

Cruciferous vegetables, spinach, rice, soy, wheat germ, wheat bran, nuts, and vegetable seeds are all included in this group.

Alkaloids

Caffeine, the best-known alkaloid, in high concentrations, is toxic and protects plants against pests and vermin and prevents the germination of all other plants in the region (allopathic effects). Caffeine is also thought to reduce the risk of diabetes and heart disease in humans and has recently been associated with the prevention of Alzheimer's and Parkinson's disease.

Spooning, in addition to other properties, spooning are called cardio protective and hepatoprotective effects. Spooning has been shown to lower blood cholestcrol levels, stimulate the immune system and inhibit the growth of cancer cells. Beans, cereals, chickpeas, oats, quinoa, asparagus, soy, red wine, and melons.

Hydroponics and accumulation of bioactive compounds

Soil farmers experience the same types of variations in soil health and fluctuating environmental conditions as hydroponic farmers. For example, water quality and variations in temperature and humidity can strain crops and possibly change their biochemical composition, regardless of the culture method used. Because of these variations, studies to date comparing the nutritional content of products grown in hydroponics with those produced in the soil have yielded mixed results. Some studies show no difference between the two methods, while others show that aboveground systems performed better or worse than those grown in the ground did. As you can imagine, the experimental design and conditions between these studies vary considerably and depend on the way they are designed and affect the outcome and meaning of the results. Several studies have claimed that hydroponic vegetables have better properties than those from conventional soil crops. On the other hand, other studies have confirmed that the exact differences between the qualities of vegetables grown in soil or hydroponics can be difficult to establish. Nevertheless, all authors generally seem to agree that hydroponic systems can be the best alternative when arable land is scarce or their types are not ideal for the desired crop. Although there are conflicting opinions, the general opinion of the researchers seems to be that hydroponics can improve the content of bioactive substances. They found an increase in macro and micronutrients and antioxidants in hydroponics, compared to soil-based production. Selma et al. found that the hydroponic system was more effective in controlling microbial contamination and showed higher antioxidant compounds, as this production method allowed for better maintenance of visual quality, tanning control, and more effectiveness in controlling microbial contamination.

The studies compared lettuce grown in soil vs. hydroponic lettuce.

Pedneault et al., in Achillea millefolium, found an accumulation of flavonoids in plants grown in hydroponics systems (0.43% dry weight) compared to plants grown in the field (0.38% dry weight). Additionally, Sherri et al. found that the hydroponic culture of basil (Osmium basilica cv. Geneva) improved the antioxidant activity of aqueous and lipid extracts, thereby increasing the content of vitamin C, vitamin E, lipoic acid, total phenols, and rosemary acid.

Crop Results Reference

The hydroponic salad offered yields 11 ± 1.7 times higher yields than conventionally produced but also required 82 ± 11 times more energy.

In lettuce, the levels of alpha-tocopherol were higher in hydroponics produced crops compared to conventional soil production.

The content of lutein, beta-carotene, violaxanthin, and neoxanthin was lower compared to land-based production, due to less exposure of hydroponics crops to the sun and temperatures, which has a significant impact on carotenogenesis by lowering their levels.

Lettuce grown in hydroponics had a considerably lower concentration of microorganisms than lettuce grown in the soil.

Total onion flavonoids were comparable between hydroponics and soil-based crops.

For red bell pepper, the carotenoid content of capsorubin and capsanthin was higher in hydroponics (4.50 and 46.74 mg/100 g dry weight, respectively) compared to conventional tillage (2.81 and 29.57 mg)/100 g dry weight, respectively).

The fruit yield per plant was 10% higher in hydroponics raised raspberries than in raspberries grown in the soil.

Sweet potato carotenes, ascorbic acid, thiamine, oxalic acid, tannic acid, chymotrypsin, and try sin inhibitors were higher with hydroponics.

There was no significant difference between the lycopene content in hydroponics and non-hydroponics tomatoes (36.15 and 36.25 mg/100 gg, respectively). Tomatoes grown in highly controlled conditions of electrical conductivity (EC), salinity of water, pH, and nutrients ensure optimum conditions for raising the levels of sugars, Bricks, pH and organic acids, quality criteria for consumer acceptance of tomatoes.

Comparative results between hydroponics and conventional soil production

Based on the results obtained by the different authors, it seems clear that they were all reported as common reasons for improving bioactive compounds in hydroponics - strict control of the entire breeding process, in particular, the amount and composition of nutrients and environmental conditions of temperature, humidity and light, and salinity of the water.

Hydroponics operations, including water recirculation systems, can provide ideal conditions for the spread of secondary metabolites, especially when plants are under osmotic or salty stress, which stimulates the natural bioactive compounds of plants. The saturation of light and temperatures by leaf receptors commonly used in these systems helps maximize photosynthesis and the subsequent production of carbohydrates that will be used for various biochemical mechanisms, including the biosynthesis of bioactive compounds, thereby increasing their content. For example, Greer and Weston found that in a controlled environment with lower temperatures, anthocyanin levels in berries increased.

Other authors have observed an increase in phenol acids, flavonoids, and anthocyanins when the ratio of day/night temperature and day/night length changed from low to high.

Another recent study with rocket salads (Eruca sativa, Eruca vesicaria, and Diplotaxis tenuifolia) found that long periods of daytime lighting (16 hours of light, 8 hours of darkness) at an intensity of 200 mol m - 2 s -1, with daytime temperatures of 20°C and nighttime temperatures of 14°C, caused an increase in the content of polyphones and glucosinolates. Similarly, day/night temperatures and day/night length, nutrient solutions, type of light, and CO_2 levels can be used to improve the bioactive compound content. Food solutions with high electrical conductivity (EC) are effective in increasing lycopene from 34% to 85% in tomato cultivar.

In a study with genera bicolor DC (a traditional vegetable from China and Southeast Asia) subjected to 80% red light and 20% blue light, supplemented with an increase in CO_2 from 450 (ambient value) to 1200 mol - 1, the content bioactive compounds has been considerably improved. The anthocyanins and flavonoids content increased from 400 to 700 mg/100 g - 1 dry weight and from 250 to 350 mg/100 g - 1 dry weight, respectively. Therefore, growing plants in a highly controlled environment can be an effective alternative to maximize the production of bioactive compounds.

Although hydroponics has many advantages over land production, several aspects should be taken into account when we decide to choose hydroponics. This type of production system requires regular irrigation and fertilization, which could lead to potential contamination of surface and groundwater in traditional farming. Also, adequate management of pH, electrical conductivity (EC), dissolved oxygen and the temperature of nutrient solutions is required, as ionic concentrations can change over time, leading to unbalanced nutrients. That is why real-time and periodic measurements of nutrient solutions are necessary and making adjustments to nutrient ratios is often necessary. Disinfection systems are also mandatory to prevent infections and diseases. All of these aspects must be taken into account to achieve high quality without compromising production efficiency and safety.

Chapter 12: Maintenance of Your Hydroponic System

Required Daily Hydroponic System Maintenance

Once a system has been designed and constructed, it is a complete, self-contained hydroponics system that will virtually run itself. There may still be some issues to address, but if you are vigilant, you will identify them early and take care of most of them easily.

Maintenance helps to support your plants as they grow. This is particularly important for an entire hydroponic system with a growing medium. It is rarely dense enough even with a growing medium to keep your larger or bigger hydroponic plants from falling over when they reach a specific size.

A plant in a soil garden can grow larger, stronger, and produce more fruit or vegetables than a comparable hydroponic plant. I actually tested this when hydroponics growing started. It was not a scientific test, but the beans that were grown in the hydroponic system produced faster and yielded far higher than the beans grown in soil.

Now back to that point. There are some essential periodic maintenance duties to perform to keep your growing system healthy in order to keep a complete hydroponics system running smoothly and efficiently. Let's start with tasks to be performed every day, or every other day if you want to.

Regular Hydroponic System Maintenance Check List

Once a developing system is up-and-running, only a few activities are needed to grow hydroponic plants effectively. Test the system every day or every other day, and keep in mind the five essential plant requirements (light, water, nutrients, temperature, and oxygen).

Some plants love humidity, and when you continuously spray them they're going to be happy.

Control the machine to make sure it works correctly. If at a given time it floods the plants and drains, you need to check that out. Small bits of growing medium can clog a system's tubing in no time flat and either leave your plants 'high and dry' or flood them continuously. This once happened to me.

As your nutrient solution evaporates, add tap water to bring the level back to where it should be. Do not apply more nutrient powder to replace what you thought was used up. This is a perfect way to kill the crop. Only the water evaporates or gets used.

Keep an eye out for rodents, disease, and nutritional deficiencies.

Make sure you take care of any possible issues as quickly as possible, or they will develop into bigger problems sooner than you think.

Dead matter saps a plant's resources and can be a start to a fungus or insect epidemic. Be sure that the dead matter stays pruned.

If you are in a greenhouse, keep track of the temperature and ventilate when necessary by opening doors, windows, and turning a fan on.

Allow some insects and breezes to get into an enclosed area like a greenhouse. This not only helps with pollination, but some bugs will actually protect the plants by killing predatory insects.

Learn to identify the good and bad bugs.

Dragonflies, mice, and daddy long legs are nice to have around—they eat the bad bugs.

Keep a log. What is second nature to you now is likely to be completely forgotten in a couple of months, so write it down.

I know that this is obviously a lot of effort, but you actually will not need anything more than a few minutes a day to execute these duties once you get a routine down. Keep the vigilance up, and you'll grow healthy hydroponic plants. You will be amply rewarded with a large number of herbs and vegetables.

Required bi-monthly maintenance

Aside from the daily maintenance criteria for your hydroponics system, there are also several essential maintenances that need to be completed twice a month or every two weeks. First, let me clarify a point about pH. It's true that the pH levels in your nutrient solution will gradually change over time and runs the risk of sliding over the plant tolerance threshold, but don't worry too much about it. Starting with a fresh, near-neutral nutrient solution (7.0 pH), you'll be fine as long as you replace the evaporated nutrient solution with water on a daily basis and perform the' 2-week task list.'

Once a hydroponics system is established, it can virtually run all by itself. However, every couple of weeks, there are a few tasks that are required.

First of all, if you have a hydroponic system with hanging roots, such as a passive or NFT growing system (and even some forms of ebb and flow), it would help to cut the roots back so that they do not clog pipes or channels through which the nutrient solution will flow. It should not affect the growth of the plant as long as you just trim a little.

Now for the toughest challenge of all: Remove the nutrient solution completely after two weeks. This needs to be done because the chemicals are not only used, but the solution's pH adjusts gradually, as it constantly flows through the pipes. It would be a bad situation for you and your plants to let it go.

To do so, unplug the electrical connection (if you have such a system) and shut off the entire system. When you grow in an environment that has no growth medium, cover growing plant's roots in a moist or wet paper towel while you are working. Dump the old solution into your soil garden or indoor plants. This product is still jam-packed with nutrients, so it can be useful.

The hydroponics system now needs to be flushed. Small amounts of mineral salts will accumulate over time as the nutrient solution passes through the system and eventually reaches a toxic level, which is hazardous to your plants.

To do this, simply hang the input and output tubes in the empty nutrient container (with the pump unattached), close the input contact valve and fill the system directly onto the growing medium, filling the entire growing medium with fresh water. Open the input valve that allows water to pour out into the nutrient container in the growing container. Dump the water and repeat the process, and the flushing is done.

Disconnect and wipe the connecting tubes, the aquarium bubbler, the pump, and the container inside out. It's going to be sticky with a bit of harmless scum, so just wipe it away. Do not use soap or any chemical, because the residue may be enough to kill your plants.

Now add tap water to a predefined level you marked off on the nutrient container, add nutrient fertilizer (my fertilizer strength is one teaspoon per gallon of water), reattach hoses, and plug it back in. This method can sound like a lengthy, labor-intensive procedure, but it is actually very easy and can be completed quickly.

You might wonder if the newly added fresh water should be a certain temperature. I didn't find that to be really important at all, and I don't pay much attention to it, other than trying to keep the new water relatively close to the old nutrient solution in temperature. I do this by touch, so I'm certainly not too technical about it and don't have any issues.

Chapter 13: Potential Problems and Solutions

Hydroponic gardens are some of the best options in terms of efficiency and how easy they are to use. However, there are a few issues that you may come across in your first few years as you learn how to get the process down.

There are several risks that you will need to watch out for because they can hinder how well your plants are growing and flourishing. Luckily, hydroponics systems are generally easy to use. It can be simple to avoid problems as long as you are aware of and looking for them. As with any hobby or new activity, there are bound to be certain problems that you will run into. Some of these are specific to hydroponic gardening while others are just general problems you may encounter. Don't worry, we have it covered!

Pests

Pests and vermin have waged centuries-old battles with farmers and gardeners and they are not about to stop with you. A list of common pests and bugs include:

- Mealybugs
- Spider mites

- Aphids
- Thrips
- Flies
- Gnats
- Whiteflies

Solution: Of course, the pests will vary from location to location but some general guidelines can be followed. Occasionally, pests will attack your hydroponic garden. When it happens, you need to use one of the following options:

Pick out the bugs and kill them by hand. This might seem somewhat repulsive but it is actually quite effective, especially if you do not want to use pesticides and the infestation is not too severe.

Make use of natural predators. As long as your microenvironment is not situated in your bedroom or living room, you might want to allow natural predators such as ladybugs or even birds and lizards to have a go at your pest problem.

Organic Pesticides

You can use organic pesticides. Here are a few tips:

- Use non-toxic pesticides, as they are the safest option for you.

- Follow the instructions provided by the manufacturers.

- Spray the pesticides in the evening to prevent the natural or artificial light from harming your plants. This will help them dry so the plants don't burn.

- If you must use pesticides, stop spraying at least one week before you harvest the plants and wash the fruits and vegetables before you eat.

Leaves

It's no secret that leaves are not only an integral part of the plant but also a warning sign for overall plant health. Some common problems and solutions seen with leaves are:

Wilted Leaves

Wilting can be caused by either too much dryness or overwatering. It can be solved simply by checking dryness. If it is too dry, add clear water (not nutrient solution!). If there is overwatering, cut down on the amount of water/solution you are using.

Drooping Leaves

Drooping leaves is usually caused by high temperatures and is a natural protective response. Also, the nutrient solution may be too concentrated. To solve this problem, simply reduce the temperature and/or check the pH of your solution with the pH meter. Flush out the old solution and add in a fresh mixture.

Yellow Discoloration

Yellowish discoloration shows the natural end to the plant life cycle. In most cases, it is normal and means it is time to uproot and start again. However, it can also be caused by white flies, in which case, follow the procedures mentioned above in the "pests" section to get rid of them. Yellow discoloration is often seen in plants where the pH is too alkaline. Dip a pH meter into your solution and check if you need to replace the solution. This should correct the problem.

Algae Growth

Algae is a natural fungus that grows in moist environments and your microenvironment with its humidity is sure to cause some algae growth. It is a little unsightly and should be removed before it gets out of hand.

Solution: Cover any ports in the nutrient reservoir with hatches, tape, or silicone to keep light out.

Use a natural fungicide like Safer Garden Fungicide.

Spindly, Thin Growth

Consider of putting the light source closer to the seedlings and try again.

Collapse

This is usually due to "rot" in the roots. It is caused by an environment that is too damp or cold. Simply increase the temperature and light source and your tiny seedlings should be fine!

System Maintenance

Remember that prevention is better than a cure. Your system will require some form of daily maintenance while your plants grow. There are many problems your system might encounter which lead to the collapse of the microenvironment.

Leakage

Your hydroponics depends on proper circulation of water and solution. Leakages will destroy your system and plants if not resolved quickly. To identify leaks, add colored water to the solution and look for the dripping of the dark color. Plug up the leak with tape or silicone immediately. Consider replacing pipes if the leak is too great.

Monitor your water pumps and humidifiers for any growth of algae or clogging by roots or leaves. They can cause spread of diseases as they supply the whole hydroponic garden.

The Lighting

You have to consider a few things, whether you are using natural or artificial lighting.

Consider These Things When Using Natural Lighting

Are your plants getting enough natural light? If you are growing inside, make sure the plant is getting enough sunlight through a window during the daytime. If growing outside, make sure the growing area is not covered by shade too much during the day.

When growing out of season, consider if the daylight is different or shorter depending on the time of year. You may have to use additional artificial lighting. With artificial lighting, you will have more control over the growing cycle and can get more harvests.

Artificial Lighting

Check to see you are using the right type of lights. Most fluorescent or LED lights are not compatible with growing plants. If needed, use HID lights.

The light should be no less than 1 foot away, but no higher than 2 feet from the plants. Adjust the light setting according to the size of your plants.

Check if your plants are too close to the lights and getting too hot in the process. Put your hand just above the plants, if it gets hot, then adjust your lights.

The Growing Climate

Your plants will not be able to grow if the growing climate is not right.

Check if the temperature of your growing system is between 60-90 degrees. Also, make sure that the nighttime temperature is 10 degrees lower than the daytime.

Check if the air is properly ventilated or not. Install a vent or, if needed, a fan so that the leaves are gently stirred around throughout the day and night.

The relative humidity of the growing area should be between 50% and 70%. Too much humidity can interfere with growing conditions and increase the growth of mold.

The Nutrient System

Check to see if there is enough solution in the reservoir. If the pump is sucking air during the cycles, then it is time to add more nutrient solution to the reservoir.

Does your solution have a balanced water-to-nutrient ratio? If you see any problems, discard the entire batch and start over. Try to keep the reservoir temperature under 80 degrees, but under 85 is also acceptable.

Check if the pH of the nutrient system is balanced. Regularly check the pH levels and adjust accordingly.

Conclusion

The hydroponic garden is not magical or enigmatic. Hydroponics is merely a way to remove or replace the usual soil with clean substitutes such as rock wool, coconut, peat, terrestrial pebbles, and so on.

In soil growing, the roots of plants will dig into and check the earth for nutrients, while in hydroponic cultivation, the exact nutrients that are designed to meet the requirements of each plant are supplied and managed automatically.

Hydroponics has proved fascinating in the field of indoor gardening. The main reasons why plant producers have adopted hydroponics are mainly economic and agricultural efficiency.

Hydroponic and sustainable hydroponic gardening will ensure that safer and healthier environmental conditions are achieved. Eventually, people who used these plant cultivation methods found it extremely enjoyable and successful. So why wait? Go ahead, launch your hydroponics garden today.

Hydroponics Gardening

How to Build Your Greenhouse Garden for Growing Organic Fruits, Vegetables, Mushrooms, and Herbs All Year Round. Learn Easy Hydroponic and Aquaponic Techniques

Scott Fields

Table of Contents

Introduction 1

Chapter 1: What is a Greenhouse garden? 7

Chapter 2: How to Build your Greenhouse Garden 13
- **Choosing a Location** 13
- **Choosing a Structure** 14
- **Choosing Covering Materials** 16
- **Constructing the Frame** 18
- **Controlling Temperature** 19
- **Additional Greenhouse Planning** 21

Chapter 3: Operation Cycle 23
- **Growing Mediums** 24
- *Coco Coir* 24
- *Gravel* 25
- *Perlite* 26
- *Vermiculite* 27
- *Rockwool* 27
- **Mixing Your Growing Medium** 28

Chapter 4: What Equipment do you Need? 31
- *The Reservoir* 31
- *Water Pump* 32
- *Timers* 32
- *Lighting* 33
- *Growth Media* 34
- *pH Test Kit* 34
- *Container* 35
- *Supporting Platform/Plant Bed* 35
- *Aggregate* 35
- *Nutrient Solution* 36
- *Aeration* 36
- *Light* 36
- *Considerations for Purchasing Hydroponics Equipment* 37
- *Available Space* 37

Chapter 5: Soil and Seeds: Getting Started 39
Germination of greenhouse seeds 39
Planting Medium 40
Containers 40
Sowing Seed 41
Temperature and Moisture 41
Maintenance of Sterile Conditions 42
Check the Temperature 42
Select a Variety of Tomatoes 43
Install an irrigation system 43
Planting 43

Chapter 6: Choose the Best Plants for Your Needs 47
Nothing tastes sweeter than fruit that you have grown yourself. Hydroponic gardening offers a great way to grow some fruit inside the comfort of your own house. Like vegetables, there are many options available to us but we'll be focusing on those that grow the best. 53

Chapter 7: Organic Farming 61

Chapter 8: Tips and Tricks to Grow Healthy Herbs, Fruits, Vegetables, and Mushrooms 71
Check the Water Quality 71
Purchase a pH Meter 72
Engage in Pest Control 72
Prepare Your Seeds for Germination 72
Place the Growing Medium Correctly 73
Select the Right Nutrients 73
Study Your Market 74
Start Small 74
Plant in Raised Beds with Rich Soil 75
Round Out the Soil in Your Beds 76
Plant Crops in Triangles Instead of Rows 76
Grow Climbing Plants to Capitalize On Space 77
Choose Compatible Pairings 77
Time Your Crops Carefully 77
Pick Fast-maturing Types 78

Stretch your season by covering the beds. Some additional tips include: 78

Chapter 9: Differences between Hydroponic and Aquaponic Methods 81
Aquaponics 81
Why choose Hydroponics for Aquaponics? 84

Chapter 10: Types of Aquaponic Systems 87
The Media Filled Bed 87
Continuous Flow 88
Flood and Drain 88
The Siphon 88
Option 1 - Timer 89
Option 2: Bell Siphon 90
Tips to Get Your System Draining Properly 92
Expanded Shale 93
Lava Rock 93
River Rock 93
Gravel 94
How to do a Vinegar Test 94

Chapter 11: How to Build an Aquaponics System – Step-by-Step Guide 95
Step 1 – Size 96
Step 2 – Growing Area 97
Step 3 – Location 98
Step 4 – Connecting It All Together 99
Step 5 – Adding Water 99
Step 6 – Cycling 100
Step 7 – Testing 101
Step 8 – Plants 104
Step 9 – Fish 104
Feeding 105

Chapter 12: Sustaining the Seasons for the Greenhouse Garden 107
1 - Move Harvested Plants and Tools 107
2 - Remove Rotten Plants and Weeds 107

    3 - Cleaning the Inside Out 108
    4 - Moving Tools Inside 108
    5 - Prepare the Soil for Spring 109
    6 - Planting Cover Crops 109
    7 - Trim Perennial Plants 110
    8 - Regenerate Compost 110
    9 - Adding Mulch to the Soil 110
    10 - Review the Growing Season 111

Chapter 13: Maintenance of Greenhouse Garden All Year Round 113

    Top Ten Reasons to Grow a Hydroponic Garden 113
    15 Tips to Make Your Greenhouse More Efficient 116

Chapter 14: Advanced Greenhouse Garden 123

    Inexpensive Ways of Cultivation 124
    Consistent Temperature 124
    Protection of Plants 125
    Increased Yield 125
    Variety and Choices 126
    Temperature Regulation 127
    Pest Free 128
    Aesthetic Appeal 128
    Multipurpose 129
    Controlling Humidity 130

Conclusion 131

Glossary 135

Introduction

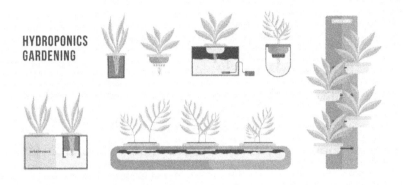

Greenhouse gardening is a way to grow plants that don't suit your environment, or a chance to start your gardening earlier in the season. Inside your home, in whatever shape or size you select, you can create an independent small ecosystem. Everything in the greenhouse affects the plants because you're working in an enclosed space, so it's a more intense way to garden.

There are plenty of kits and full greenhouses available from a wide range of suppliers. Gardening supply companies, lumber yards, hardware stores, and most large chain stores have stocks at their disposal. If you're so inclined to create your own greenhouse, there's a lot to be said for that. To simplify the task of this do-it-yourself project, buy only the frame of a garden shed package. It saves you from building the rafters and framing a doorway, so it speeds up the process. From there, you can use your choice of materials and design your ventilation options according to your preferences.

You can develop the greenhouse ecosystem with normal planting soil, water, sunlight, and fresh air components. The soil must be rich in fiber so that seeds or seedlings can easily set roots. Well-rotted manure is a good starting point if it's available. You must water everything daily, whether it rains or not.

In the greenhouse, fresh air or air circulation is important for successful gardening, and you can provide it in two ways. The air circulation may be adequate if you position the greenhouse to take advantage of natural winds and have adequate ventilation openings such as windows. Use one or more fans to increase ventilation as needed.

If the temperature and humidity spike throughout the hot summer days, your gardening experience will not be successful. You need to track this information so you know what's going on. You can track the temperature and humidity with an inexpensive monitoring system that enables a greenhouse sensor to transmit to your household receiver. This device also helps you know when you need to protect from frost. A small heater will do the job.

Gardening has three aspects. The first is inside your home for your garden. The second is to use outdoor space for planting your garden. The third is under glass, which we call greenhouse gardening.

Greenhouse planting is similar to gardening outside, but you must control the temperature of the greenhouse. Keep in mind that plants do much better at temperatures that are slightly below house temperatures and need more humidity. For your greenhouse plants, this will be the perfect environment.

You need to build your greenhouse in a location that maximizes the amount of sun it can get throughout the year. This is important when the sun is at its lowest point for the spring and fall. Locate your greenhouse in a southeast to a southwest direction where the sun will be.

To get better ventilation flow, space your plants around the greenhouse area. In the morning, open the greenhouse doors and then close them in the late afternoon. You can also do this in the winter as long as you watch the weather and keep the inside from getting too cold.

Greenhouses contribute greatly to any garden. We encourage you to grow varieties of plants that are not appropriate for outdoor areas and expand the growing season to other crops.

It may seem daunting to know where to start for novices setting up a greenhouse.

Heat: On sunny days, the sun's rays will deflect through the window, and the greenhouse will heat up beautifully. But in colder weather, you will need other forms of heating. Electric heaters are the simplest to install, but you can also use gas (although this approach could cause fumes in your greenhouse).

You will also need a way to prevent the temperature from becoming too high in the greenhouse. Do not add air conditioners for ventilation because they will dry the air. Natural wind or exhaust fans will do the job. You also need soil that drains enough to prevent water logging but maintains moisture so the plant roots can absorb the water. Ideally, the soil should be slightly acidic and contain a lot of organic material, which is essential. The only way to do this is to add compost.

Compost: You need to apply a good proportion of compost to your growing beds because the soil will not have any organic matter, the rich topsoil full of nutrients and bacterial activity. For example, leaf litter, animal droppings, and rotting plant material will be a good compost mix. Many guidelines suggest that you apply compost to a third of your beds. This will provide your plants with the best nutrients to grow and survive. If possible, choose an organic fertilizer or, even better, create your compost pile with scraps from the kitchen and plant cuttings. A greenhouse is an environmental enhancement and strategic planning system that allows the cultivation of plants in climates that would not otherwise be well suited for their growth.

You can use greenhouses to cultivate several types of plants with a wide range of environmental conditions, from rainforest plants to succulents and anything in between.

Chapter 1: What is a Greenhouse garden?

A greenhouse is a structure whose walls and roof are mainly composed of transparent materials, such as glass, in which plants are grown in controlled climatic conditions. The size of these structures ranges from small hangars to industrial buildings. A miniature greenhouse is known as a cold frame. The inside of a sun-exposed greenhouse becomes much warmer than the outside ambient temperature, thus protecting its contents in cold climates.

Many glasshouses or greenhouses are high-tech structures to produce vegetables or flowers. Glass greenhouses have equipment such as filters, heating, cooling, and lighting, and you can control it by a computer to optimize plant growth conditions. Furthermore, you can use various techniques to evaluate the degree of optimization and comfort ratio of the greenhouse microclimate (air temperature, relative humidity, and vapor pressure deficit) to reduce production risks.

- In the seventeenth century, they constructed a greenhouse with ordinary brick or wood, with a normal proportion of windows and some means of heating. When glass became available, and heating forms became more sophisticated, the greenhouse became a glass structure. A considerable increase in the availability of exotic plants in the 19th century led to a strong growth of greenhouses in England and elsewhere. Major industrial greenhouses play an important role in agriculture, horticulture, and botany, while amateurs and home gardeners often use smaller structures.

A modern greenhouse is usually a glass or plastic structure to manage vegetables, fruit, flowers, and many plants that require special climate conditions. The basic structural forms are narrow with an A-shaped sloped roof. Sometimes individuals join two or more greenhouses, so there are fewer exterior walls, which reduces heating costs.

Greenhouses have large glass areas on both the sides and the roof to expose plants to natural light most of the day. Glass is the traditional material, but you can use plastic films, such as polyethylene or polyvinyl, and fiberglass also. The greenhouse is partly warmed by the sun's rays and partly by artificial means, such as steam, hot water, or circulating hot air. As a greenhouse can become too hot and too cold, a type of ventilation system is also needed; they are usually roof openings (mechanically or automatically), and electric fans to attract and circulate air throughout the interior.

Plants grown in a greenhouse come in several broad categories depending on their overnight temperature requirements.

In a cool greenhouse, the night temperature drops to about 7 to 10°C (45 to 50°F). Azaleas, cineraria, cyclamen, carnations, fuchsias, geraniums, peas, guillemots and various bulbous plants, including daffodils, lilies, tulips, hyacinths, and daffodils are among the plants suitable for a cold greenhouse.

The standard greenhouse has night temperatures from 10 to 13°C (50 to 55°F). Begonias, gloxinias, African violets, chrysanthemums, orchids, roses, teapots, and various types of ferns, cacti, and other succulent plants adapt well to these temperatures.

In a tropical greenhouse with a nighttime temperature of 16-21°C (60-70°F), you can grow caladiums, philodendrons, gardenias, Easter flowers, bougainvillea, passion flowers, and various types of palms and orchids.

In cold weather countries, the commercial greenhouse is used to grow tomatoes and other hot weather vegetables. A greenhouse is a building with glass sides and a glass roof in which plants need to be protected from weather or environmental sensitivity. Greenhouses are different from typical farms, which consist of areas where grains or animals are grown. Growing plants in a greenhouse can be your ultimate dream; however, there is more to a greenhouse than just setting up the structure. You need to know how to maintain the greenhouse and ensure there are optimum conditions for the plants to thrive.

Having your own greenhouse will enable you to grow vegetables all year round, exotic plants, or herbs, and start the seeding process early. There are many reasons why you should set up your own greenhouse garden.

Each greenhouse model includes some temperature control features and other components to help you utilize the greenhouse's functions. Some of these components or amenities include electricity, heat, water, lighting, shelves, and benches. For example, the heating system enables you to grow your plants anytime of the year, so you don't have to worry about which season is the best for growing each crop. The lighting will enable you to walk into the garden, even in the dark of night, and work on your crops, including planting new ones, trimming, and cutting.

Greenhouses allow temperature and humidity to be adjusted in enclosed spaces so that certain crops can grow and thrive regardless of the weather outside. A greenhouse is an environmental enhancement and strategic planning system that allows the cultivation of plants in climates and seasons that would not otherwise be well suited for their growth.

Chapter 2: How to Build your Greenhouse Garden

Choosing a Location

1. Select an area facing south or north (depending on your location). The main element needed for the greenhouse is good sunlight for as many hours per day as possible.

All structures must be north of the greenhouse. One of the main structures of the greenhouse is a shed. Choosing a south wall of a building is a good option.

2. Assign prefcrences to places where the morning sun is higher than the afternoon sun. Although the best option is the sun all day, opening the area in the morning light will accelerate the growth of plants.

If there are trees or shrubs near the greenhouse, make sure they do not shade the area until late afternoon.

3. Watch out for the winter sun and the summer sun. If the area to the east is open and sunny, there will be more sun from November to February.

Do not choose a place near to evergreen trees. Deciduous trees lose their leaves and will not shade the site in winter when the greenhouse needs more sun.

4. Choose a place that has access to electricity. Most greenhouses require heat and ventilation to maintain optimum temperature.

If you build a shelter, you can extend the power from the home.

A separate building may need work to be done by an electrician.

5. Choose a well-drained space. You have to evacuate excess rainwater away from the greenhouse.

If your locality is uneven, you may need to fill the area to promote drainage.

You may be able to use tanks to collect rainwater that flows down the gutters of your greenhouse. Any conservation of water and electricity will help keep greenhouse costs down.

Choosing a Structure
1. Measure your position for placement of the greenhouse. Whether you are building the greenhouse from scratch or with a kit, you must choose the correct size.

The larger the greenhouse, the more expensive the construction and heating will be.

The most effective greenhouse size is 2.4 x 1.8 m (8 x 6 ft).

2. Choose a greenhouse kit if you have little construction experience.

You can get an emerging greenhouse or polycarbonate greenhouse at a home improvement store. Larger and more robust models range from $500 to $5,000, depending on their size.

3. Make a well. If you have chosen a space against a building, you can also build a simple submissive structure based on the remaining wall.

If you have a brick structure, the heat of the building can help you maintain a warm and stable temperature.

It's a fairly easy structure to make. You can support with rebar, wooden beams, and fewer supports than a complete construction project requires.

4. Build a Quonset frame. This is a vaulted roof that can be made with steel supports or PVC tubes (PVC contains many imitators of water-soluble carcinogens; lope tubes are a good alternative but more expensive).

The shape of a dome model means that there is less headroom and storage than the rectangular models.

This form can be built at a low cost. However, the less expensive the material, the less likely it is to be solid.

5. Choose a rigid box. With this concept, you need a base and a frame. Unless you are a designer, you will want to buy a greenhouse plan or employ someone to build it.

A rigid frame, a pole and a bundle, or a saw with frame style A requires a sturdy base and frame.

You will need the help of friends or employees to help you build a large frame greenhouse.

Choosing Covering Materials
1. Use a UV stabilized polyethylene that is economical but contains BPA or LDPE, which is more expensive but not toxic and lasts longer. It is stabilized against UV rays. It should be washed occasionally.

2. Use dual-wall hard plastic, such as polycarbonate or multi-wall corrugated polycarbonate, or acrylic (Plexiglas), which is more expensive, but does not contain any gaps, which offers greater transparency to light.

Polycarbonate can be slightly curved around the frame and can save up to 30% energy because it has a double wall.

Polycarbonate is 200 times stronger than glass. The polycarbonate also has a high light transmission and is stabilized against UV rays, but contains water-soluble BPA. Acrylic is light, but not so strong (stronger than the glass). About 80% of the light is allowed through the polycarbonate material. This is 90% true for acrylic.

3. You can get fiberglass by building a frame saw, but this time you have to save money for the fiberglass instead of glass because roof construction can be lighter than construction. The fiberglass turns yellow and loses its transparency in a few years. Acrylic is more expensive but more transparent and remains transparent for up to 10 years.

Choose transparent or even better acrylic glass fiber. A new layer of resin will be needed every 10 to 15 years. You can also invest in high-quality fiberglass. Light transmission is greatly reduced in lower quality fiberglass. By purchasing acrylic, this will save you about 25% of the cost compared to glass and allows you to work with it more easily.

4. Choose the most attractive material if you are building a greenhouse that will highlight your home or garden.

Glass is very fragile and replacement in case of breakage is expensive, but on the other hand, acrylic, fiberglass, and polycarbonate need to be replaced in time just from aging of the materials.

You have to build a greenhouse with a base; any misalignment due to salvage can cause harm.

Tempered glass is preferable because it is stronger than ordinary glass. Think about using tempered glass for the roof. A 4 mm glass thickness is recommended if it is in an area prone to hail.

If you plan to install glass at your company's expense, you should consider using a construction company to ensure that the base and frame can support the weight.

Constructing the Frame
1. Use chains along the plane to determine where you want to put the brackets.

2. Reinforce with rebar. If you are building a cottage or Quonset, you can reinforce your frame with straps and PVC or a non-toxic variant.

Place the bars against the ground every 1.2 m (4 feet). Once the frame is fitted, you can make a 20-foot tube section to create your frame. Tighten your plastic film (preferably a non-toxic type of plastic) into the frame and secure it to the bundles at the bottom.

3. Pour the gravel on the ground in a uniform layer, once its supports are inserted into the ground. This also allows additional drainage in the greenhouse environment.

Sink the greenhouse floor before it is framed.

4. Treat the wood before use, but consider what you are using for treatment, as not all coatings and treatments are suitable for contact with foodstuffs.

Untreated wood can degrade in just three years.

Carefully choose your wood treatment. Some wood treatments require that foods are no longer listed as "organic" or safe for consumption.

Consider a treatment like Erdalith, whose smoothing properties are limited.

Use metal racks instead of wood racks when possible.

5. Attach the liner to the frame as close as possible.

Look for the best procedure for coverage of your choice.

Controlling Temperature

1. Put fans in the corners of the greenhouse. Set the fans diagonally to create airflow.

They have to run almost constantly during the winter months to ensure that the entire greenhouse benefits from heating.

2. Install fans on the greenhouse roof. They can also be located near the top of the support beams.

Some decomposition of carbon dioxide is essential.

The airflow can be adjustable. You need to open doors and windows more widely during the summer months.

3. Solar heat can account for only 25% of the heat in the greenhouse. Therefore, an additional electric heater is essential.

You can also use a wood appliance or heating oil, but it must be ventilated outside to ensure good air quality. Carbon dioxide poisoning is a real danger in an enclosed space.

Check with your city or town council to find out what heating options are allowed in your area.

4. Install a forced-air system, if using glass. If you can afford to equip your greenhouse with your own temperature control system, you can configure it to grow almost anything.

Use an electrician and a contractor to install your system.

You may need regular maintenance to control ventilation and heating in winter.

Located in different levels of the greenhouse, you can observe the temperature in the greenhouse at any time.

You can buy a thermometer that measures the temperature so you can watch it closely during the winter months.

Additional Greenhouse Planning

1. Study the planting conditions of the plants you want to grow. The more sensitive the plant is to temperature changes, the less it is possible to grow other plants in the same section.

A cool house is a greenhouse designed to prevent plants from freezing. It is ideal for temporary greenhouses.

You have to choose the temperature and keep it stable. It is not possible to create different temperature zones in an open greenhouse.

2. Make sure you have a constant water supply. Ideally, it should be powered by a garden handle and tanks.

3. Construct elevated beds in the greenhouse. Slat panels can be used, as they allow water to flow through the table and into the gravel floor.

If possible, build the upper beds to benefit the main gardener, to limit ergonomic problems.

Chapter 3: Operation Cycle

Now that we have a hydroponic system set up, let us take some time to look at how the operation works. This means that we will be exploring the different kinds of growing mediums available to us to see what works best for which kinds of setups. We will also explore how we seed our hydroponic gardens, how we light them, and what we do when the time comes for trimming.

Growing Mediums

When it comes to what medium we use in our grow trays, there is a ton of variety available to us. This can be a little intimidating at first when you aren't sure which medium is right for you and the gardening that you are looking to do. It is important that we choose a medium that works with the plants we are planning to plant. This means that we have to take into account things like water retention and pH balance.

Before we look at the mediums themselves, a quick word on the requirements of the different systems. The way that each system is set up and works actually says a lot about what kind of growing medium works best. For example, a drip system functions best when it is using a growing medium that doesn't become too soggy. In contrast, a wick system likes a growing medium that absorbs and holds onto water and moisture with ease. With nutrient film technique systems, you want to avoid a growing medium that easily saturates. An ebb and flow system will want to have good drainage and a growing medium that doesn't float. Considering the mechanics of your system of choice is the first step in deciding a growing medium.

Coco Coir
An organic and inert grow medium, coco coir is made from the frayed and ground husks of coconuts. When it comes to pH, coco coir is very close to neutral. Coco coir retains water but

also allows a decent amount of oxygen to get through which helps the roots. This medium is primarily used in container growing or in hydroponic systems of the passive variety such as wicking. Because it can clog up pumps and drippers, it is not a great choice for more active systems such as the ebb and flow system.

Gravel

Gravel doesn't absorb or retain moisture. Instead, gravel works to give an anchor for the roots of the plant. For this reason, gravel works best in a system that doesn't require a ton of retention such as a drip system or a nutrient film technique system. Any system that keeps the roots of the plant in constant contact with the water can make good use of gravel.

In some setups, such as the bucket-based drip system we saw above, gravel is used as a bottom layer in the pot. This allows for better drainage as the water has passed through whatever medium made up the top layer to find gravel, which doesn't retain it whatsoever. It also serves to add some weight to the bottom of your tray to prevent spills from wind or other elements.

If you are using gravel, make sure to give it a proper wash before using it in the system. If you want to reuse the gravel, make sure to wash it yet again. We do this to prevent salts or bacteria from getting into the hydroponic system and causing issues such as burnt roots, high levels of toxicity, and the like. Jagged gravel can also damage the roots so it is best to use smooth gravel as a way of avoiding this.

Perlite
Perlite is actually an amendment to our growing mediums, which means that it is used to improve an existing medium rather than just being used on its own. You make perlite by heating up glass or quartz sand, though of course we don't have to make it ourselves. You can buy it from any gardening store. Perlite helps improve drainage and aeration when mixed with another growing medium such as coco coir.

Because we are using a nutrient mix and not just pure water, we have to be concerned about nutrient build-up. The nutrients in our solutions can get absorbed into the growth medium and lead to a build-up of toxicity, which risks killing off our plants. No gardener wants that. The extra drainage that perlite offers will help prevent this build-up, making sure that our plant's root system gets the oxygen it needs to grow. Perlite comes in different grades from fine, and medium, to coarse. The kind you need will be determined by your potting mix. Perlite should never take up more than a third of your mix. Using too much will cause it to float and floating perlite doesn't offer the benefits we wanted it for in the first place.

Vermiculite
Vermiculite is actually a lot like perlite. It comes in three different grades, again being fine, medium, and coarse. Made by expanding mica through heat, vermiculite is another soil and potting mix amendment. This means that vermiculite is mixed with another growing medium in order to get the best results.

Vermiculite works opposite of perlite. Where perlite helps with drainage of our growing medium, vermiculite helps our growing medium to retain water. For this reason, vermiculite is often mixed with perlite for use in hydroponic systems of the passive variety such as wicking systems.

Rockwool
One of the most popular of the growing mediums, Rockwool is made through the heating and spinning of certain silica-based rock into a cotton candy-like material. This creates a firm material that tends to have the ideal ratio of water to oxygen that our plants' roots love. It also is mostly pH neutral, which is always a plus.

It can be found in a bunch of different shapes and sizes with the most common being a cube shape. These cubes are awesome for starting out seeds (which we'll look at more in just a moment). These smaller cubes are often used to begin growing plants.

Because of the versatility of Rockwool, it can be used for starting plants before transferring into another medium for deep water cultures or nutrient film technique systems. It can also be used for drip systems and ebb and flow systems without the need to transfer.

Mixing Your Growing Medium
When it comes to which growing medium is the best, it depends on the job that you are trying to tackle. Once you have an idea of what you need, you can begin the task of mixing it all together. There are many different projects on the market that offer pre-mixed growing mediums and these can be a great way to save a little time and get what you need right out of the box.

However, some of us like to get our hands dirty in this part of the process. Mixing your own growing medium can be a great way to make sure it is 100% the way you want it to be. This can be a little tricky if you are new to hydroponic gardening and don't know what combination of mediums is best. Part of getting into anything new, and hydroponic gardening is no different, is that you have to accept some uncomfortable moments and you have to accept that you will learn from your mistakes.

For an example of one mixture, let us look at what Upstartfarmers.com has laid out in their discussion on soilless potting. They offer a formula for a mixture that is one-part coconut coir or peat, one-part perlite or vermiculite, and two parts compost. While the systems we have looked at aren't focused on compost but rather getting nutrients through our reservoir's solution, this shows us a straightforward mixture. Notice that the perlite or vermiculite does not exceed 33% (or 1/3rd) of the total mixture.

Chapter 4: What Equipment do you Need?

There are several hydroponic kits available for purchase. If you want to build a custom system you need to know what elements make up an effective setup.

The Reservoir
The reservoir utilized in hydroponic structures holds the water that contains the nutrients to be supplied on your plants. Relying on your finances as well as the scale of your operation, the reservoir may be something from a pricey

business variation or as easy as a simple bucket. Make sure to choose a container that includes a lid. Moreover, a satisfactory reservoir should not be metal as this can result in the advent of harmful minerals into the nutrient solution or the incidence of chemical reactions that can damage your plants.

Water Pump
To deliver your plants with the water and minerals they need to live on, you need to search out a dependable water pump. The two essential varieties of water pumps are submersible and non-submersible. As you might have guessed, the submersible is installed inside the nutrient solution. The non-submersible desires to be connected to the system outside of the solution. Water pumps are categorized consistent with their output in gallons per minute (GPM) or gallons per hour (GPH). When you have a small installation, then a pump that supplies around 30 to 40 GPH should be able to provide your plants with the water they need.

Make sure to consider the rate at which water drains from the grow media when deciding on a water pump to meet the desired degree of output.

Timers
In maximum hydroponic structures, except the most fundamental ones, a timer is required to help with the regulation of several essential processes. For example, a timer may be used to adjust watering, ventilation, and lighting

cycles. While deciding on a high-quality timer for your system, you may have choices between basic models that are more affordable analog gadgets or costlier superior virtual devices. The more expensive models will be first-rate for developing sensitive plant life that requires utmost accuracy for the execution of every step in the operation.

Lighting
To decorate the boom of your plants, you want to have the proper lighting. Although fluorescent lighting may be used to supplement natural light, they cannot provide the spectrum of light wanted by most plants. Steel halide and sodium lighting will emit a spectrum of light that mimics the light emanating from the sun. Metal halide lighting fixtures are the closest you can get to natural daylight. They produce a higher percentage of blue light that is remarkable for assisting vegetative growth. Excessive strain sodium lights produce light that covers a greater portion of the pink-orange spectrum. They burn longer and brighter and use a lower quantity of energy than steel halide lights, even though they produce a narrower spectrum of light. For the best results, it is recommended that you combine both kinds of lighting fixtures to provide light this is as near as possible to the entire spectrum of daylight. Furthermore, you can use light reflectors and movers to cover a much wider area with fewer lighting fixtures.

Growth Media

The medium chosen should be able to anchor the plant, facilitate proper drainage, and provide aeration of the roots. The proper medium should be sufficient to anchor the plant; however, not so much that it hinders circulation of air and use of the nutrient solution. The particles of the medium ought to be able to preserve moisture and nutrients long enough to permit the roots to absorb the necessary level of nutrients. Additionally, it has to be sterile to save you the risk of plant sicknesses, pests, and parasites.

pH Test Kit

You want to maintain the pH stability of the nutrient solution to have any chance of growing a wholesome hydroponic garden. Although a few plants can thrive at a different pH level, it's suggested that you keep the pH between 6 and 6.5. You may be wondering if you need to buy a pH test kit. Of all the hydroponic gadgets discussed, the all-in-one kits are usually the least expensive, but also have the most vital tools. Growing a hydroponic garden involves less work than growing in the soil. However, to prevail, you want to have the most helpful gadgets, such as pH meters, from the beginning, irrespective of whether you choose a ready-made kit or are planning on building your own system step by step.

Container
No matter the technique used, all hydroponic gardens require some form of box or tank to hold the nutrient solution that feeds the plants. Many homemade hydroponic gardens use a huge storage container, water trough, or even a kiddie's pool for a tank. The material used to make the tank isn't critical as long as it's watertight and blocks out most of the light. A great rule of thumb is to have a box that is between 6 and 18 inches deep and 2 – 5 feet across.

Supporting Platform/Plant Bed
A platform or plant bed is essential to help the plants as they develop. For systems where the roots grow inside the water, the plant life is positioned in holes in the platform and held in place with cotton or comparable cloth packed across. For mixed-culture systems, the vegetation is positioned in a non-soil growing media, also referred to as combination or muddle. The aggregate may be held in mesh containers that can be placed within the supporting platform, or the platform may be a trough with a porous bottom. This is packed with aggregate. The nutrient solution is carried out so it is in contact with the roots.

Aggregate
A broad type of substance can be used in a hydroponic combination. These include sand, gravel, vermiculite, timber shavings, peat moss, perlite, and a fibrous stone fabric

referred to as Rockwool. The only requirement is that is does not offer any nutrients to the plant. It should also be strong to appropriately guide the plants as they grow and be layered at least 3 inches.

Nutrient Solution
The nutrients needed for a plant to grow in a hydroponic device has to be provided via the nutrient solution. The kinds of nutrients that need to be present in the solution are nitrogen, phosphorus, potassium, and calcium. It's far more convenient to buy a premade solution for general and effective use than to try to prepare homemade solutions.

Aeration
Beyond food, the plants need to draw oxygen into their roots. Aeration is vital to make sure there is sufficient oxygen within the nutrient solution. For smaller gardens, this can be performed through emptying and refilling the container with solution. Larger systems regularly use an air bubbler, similar to those in aquariums, to keep the nutrient solution oxygenated.

Light
Hydroponic gardens may be grown inside or outside. Because vegetation, particularly greens, need lots of light to thrive, an artificial light source can be ideal. Fluorescent or incandescent

bulbs may be used as long as they offer enough light from the pink and blue range of the visible light spectrum.

Considerations for Purchasing Hydroponics Equipment
Coming into the arena of hydroponics can be very fun and enjoyable; however, it could also be extremely intimidating. There are so many options for hydroponics that it can be overwhelming. At the same time, as it can be tempting to buy the first inexpensive gadget you find, you don't want to wind up with a water system that doesn't fit your needs. Here are the most critical factors to consider while buying your system.

Available Space
Where will you grow your plants? A small greenhouse outdoors? A large closet? Your basement? Before you purchase device system, make sure you calculate the space you will be using and figure out exactly what size of hydroponics system you can install there. In case you are planning on developing rows of plants, attempt to allow at least one meter of walking area between each row to make it simpler to tend to the plants.

Your Plants

You probably already have an idea of what equipment you need to grow hydroponically. The choices of plant life are more extensive. Spend your money on numerous 18-inch buckets if all you need to grow is smaller plant life. Talk to your hydroponics retailer about what form of device, medium, and fertilizer will accommodate the dimensions, types, and amounts of plant life you want to grow. Many producers will allow you to speak to hydroponics professionals about these types of growing troubles.

Your Budget

Before you buy your hydroponics system, you should decide how much money you are willing to spend. Start-up costs aren't the only expense associated with hydroponics. You need to factor in power costs for your lights and how often you will have to update your system. If you plan on maintaining your hydroponics system for years, it could be less expensive in the long run to purchase the system.

Your Time

You probably don't need to devote all your time to growing your plants. An aeroponics gadget might be an attractive option. However, if something goes wrong with the timer, the roots may dry out. These sorts of systems can sometimes require extra time and attention. Look for a machine that offers you a more considerable margin of error, inclusive of one that incorporates a medium that holds a good amount of air and water.

Chapter 5: Soil and Seeds: Getting Started

Germination of greenhouse seeds

For the gardener, nothing gives us as much hope as a well-functioning greenhouse. There is always the possibility of a new life born of these small, lifeless fragments we call seeds. Add some water, heat, a good mix of soil and, there you have it, a new life. If only it were that simple. The greenhouse provides the ideal environment for plant growth, but it also provides the ideal environment for disease proliferation. With care, a little knowledge, and the right tools, spring can spend all year in your greenhouse.

Planting Medium

A good germination medium for greenhouse seed is usually composed of a combination of vermiculite, perlite, peat, coarse sand, treated bark, or expanded shale. However, there are commercial mixes that offer several advantages over domestic media. These mixtures are certified free of weeds, insects, and diseases. They are convenient and ready to use directly from the bag and already contain a small amount of fertilizer incorporated into the mixture to keep the seedlings for 2-4 weeks. The smaller the seeds for planting, the better the soil should be.

Containers

Containers used for seed germination in greenhouses should be sterile. Constant high humidity and greenhouse heat provide ideal conditions for the development of plant diseases. Flatbeds, 2 to 3 inches deep, with holes in the bottom for drainage are the best type of container for most seeds such as beans or grain. The seedlings will grow rapidly and absorb excess water.

Sowing Seed

The container is filled with germination media up to 1/2-inch from the top. The surface is moistened and the excess water is evacuated. The seeds are evenly distributed on the top of the medium. Planting depth is indicated on most seed packets. Seeds sown too deep cause congestion and little growth. Tiny seeds, such as petunias or impatiens, are left untouched or covered with a very thin layer of vermiculite. The seed fragments are covered with coarse sand or sphagnum vermiculite. Water the new plant taking care not to disturb the surface material layer.

Temperature and Moisture

It is best to keep seedlings at the temperature recommended for the species, indicated on the seed package. Most seeds germinate at temperatures between 70 and 80°F in 7 to 10 days. To accelerate germination, a heating cable, available at the garden center, is set at a temperature of 70 to 85°F and placed under jars with seeds. High humidity is required for seed germination, preferably 100%. Manual irrigation is best for small batches of pots with seeds, being careful not to disturb the surface of the soil. For large plantings, use a sprinkler with a flow tube.

Maintenance of Sterile Conditions

The sterility of the containers, supports, and tools used for seed germination is the most important factor for successful growth. New seedlings are particularly susceptible to plant diseases, such as fungi and bacteria, which can persist for a long time in pots, soil, reservoirs, and tools, and then kill a complete planting in a few days. Containers must be new or disinfected in a 10% bleach solution for 5 minutes prior to use. Usually a sterile mix is your best option and the habitat should contain only sterile components. All instruments or tools that touch the floor can be easily sterilized in a 10% bleach solution. The benches and greenhouse boards should be disinfected in the same way that the trays and tools to ensure sufficient airflow between the trays.

Check the Temperature

Tomatoes grow best at day temperatures of 21 to 27°C (70 to 80°F) and 16 to 18°C (60 to 65°F) at night. Be sure to maintain these temperatures in the greenhouse for several months before planting.

Ideally, set the temperatures at the low end of this range on rainy days and increase them to a maximum (or even a little higher) on sunny and bright days.

Select a Variety of Tomatoes

There are thousands of varieties of tomatoes, so it is best to contact local producers for detailed information. However, some guidelines and tips apply to all regions: If you run out of space, plant a "certain" variety that stops at a certain height.

Choose a traditional tomato that can grow on any well-drained material.

Pearl wool or rock wool bags are the smallest options in many areas. Some manufacturers prefer a 1:1 mixture of peat and vermicelli.

Buy some sterile soil or make your own. Never use soil or compost from your yard or garden without sterilization. Choose this option if you do not want to install an irrigation system.

Install an irrigation system

Install drip tubes to supply water to each plant. A tube-connected fertilizer injector can also automate fertilization. Tomatoes are easy to grow in a hydroponic system.

Planting

1. Fill a starter tank with dirt. Wash the pan with soap and water to disinfect it. Fill the pot with one of the potting mixes

described above. If using the soil from the ground, make sure it is sterile.

If you use a soilless mixture, you also need a good nutrient solution.

2. Drill a 6 mm (1/4 in) hole in each container. Plant one seed in each hole. Lightly cover with compost soil. Plant 10-15% more seeds than you want. Later you will eliminate the weakest seedlings.

3. Moisten with water or nutrient solution. Use simple ground water or seedling nutrient solution for soilless mixes. In both cases, water until the mixture is moist enough to form a pool, with a few drops drained. Water regularly to keep the mixture moist.

A 5:2:5 nutrient solution containing calcium and magnesium is ideal. Dilute the solution according to the instructions on the label.

4. Keep the containers in front of a warm window. Do not bring seeds into the greenhouse until they have germinated to avoid diseases and parasites. Make sure there is enough sunlight and maintain the temperature between 24 and 27°C (75 and 80°F) throughout the day.

Move them to full sun once all the seedlings are growing. This usually lasts 5 to 12 days.

5. Transplant into larger containers. Transplant the plants in small pots in a greenhouse approximately two weeks after emergence. After six to eight weeks, or once the plants reach 10 to 15 cm (4 to 6 inches), transplant larger plants into pots or bags. A typical plant needs about 1 to 2 cubic meters of space (3.7 to 7.5 gallons or 14 to 28 liters). Some varieties will produce less fruit if they are grown in small pots.

If you see insects, molds, or disease spots on a plant, do not bring them into the greenhouse. You risk contaminating the entire crop.

6. Adjust the levels. Before the final transplant, you may want to check and adjust the pH of the soil, which ideally ranges between 5.8 and 6.8. If your soil is too acidic, add about 1 tablespoon (5 ml) of hydrated lime per gallon (3.8 L) of soil. In addition to increasing the pH, add calcium to prevent the rotting of flowers later. If not, choose only calcium-containing fertilizer and apply every week or two.

Chapter 6: Choose the Best Plants for Your Needs

We know what each of the hydroponic garden setups are, how we can make our own, and what kind of operation cycle we can expect. In this chapter, we are going to take a look at the different plants that are available for us to grow. We will take a brief look at each plant to get an idea of how they will best grow in our hydroponic setups. From there, we will be looking at the nutrition that our plants require.

Vegetables

When it comes to vegetables, there are many options available to us. We'll be looking at a handful of these but first, let's tackle some general rules of thumb.

First up are those vegetables that grow beneath the soil. These are vegetables like onions, carrots, and potatoes. These plants can still be grown in a hydroponic system but they require extra work compared to those that grow above the surface like lettuce, cabbage, and beans. This means that those under-the-soil plants require a little more advanced skill, and you may want to get some experience with your hydroponic system before you try to tackle them.

The other rule of thumb is that we should at first try to avoid crops like corn, zucchini, and anything else that relies on growing lots of vines. These types of plants take up a ton of space and just aren't very practical crops for hydroponic systems. Instead of focusing on a plant type that isn't practical, we can make better use of our space and systems for higher yielding crops.

Beans

There are many different types of beans from green beans to pole beans, lima beans to pinto beans. Depending on the type of bean you plant, you may want to consider adding a trellis to your setup. Beans offer a wide variety for what you can add them to and they make a great side dish to just about any meal. When it comes to temperature, beans prefer a warm area. They also prefer a pH level of around 6.0.

If you are growing your beans from seeds, you can expect them to take between three and eight days to germinate. From there you can expect another six to eight weeks before it is time to harvest. After harvesting begins, the crop can be continued for another three or four months.

Cucumbers

Like beans, there are a few different options when it comes to what kinds of cucumbers we can grow. There are thick-skinned American slicers, smooth-skinned Lebanese cucumbers, and seedless European cucumbers. There is a wide variety and the best news is that they all grow pretty well in a hydroponic setup. Where beans prefer a warm temperature, cucumbers prefer hot. They like to be a step beyond just warm. They also prefer a pH level between 5.5 and 6.0

It only takes between three and ten days for cucumbers to begin to germinate. They take between eight to ten weeks to get ready for harvesting. When it comes to harvesting cucumbers, make sure that the cucumbers have taken on a dark green color and that they are firm when you grasp them. Because each cucumber grows at a different rate, you can expect the harvesting to take some time, as you don't want to pick them before they are ready.

Kale

Kale is a delicious and nutritious vegetable that makes a great addition to just about any meal. There are so many health benefits to kale that it is often considered a superfood. Kale actually prefers a slightly cooler temperature; it grows best in a range from cool to warm. Like cucumbers, kale prefers a 5.5 to 6.0 pH level.

From seed to germination only takes four to seven days. However, harvesting takes between nine and eleven weeks. It's a little bit longer to grow kale than beans or cucumbers but you can harvest it in such a way so that it continues to grow. If you only harvest 30% of your kale when it comes time, this allows it to quickly regrow. Doing this means that you can easily keep this super food in your garden and in your diet for several months.

Lettuce

As you have been reading through this book, I would bet it's safe to say that no plant has popped up more often in our discussion than lettuce. This is because lettuce absolutely thrives in hydroponic growing conditions, which is great since lettuce can be used to make salads, add some texture and flavor to our sandwiches and burgers, and is just an all-round versatile vegetable to have in the kitchen.

Growing lettuce offers a lot of variety. While lettuce prefers a cool temperature and a pH level between 6.0 and 7.0, it works in any hydroponic system that you make. For this reason, lettuce makes a great entry plant for getting into hydroponics. Lettuce only takes a couple of days to germinate but the time to harvest depends on what kind of lettuce you decide to grow. For example, loose-leaf lettuce only takes forty-five to fifty days to harvest. Romaine lettuce can take up to eighty-five days.

Peppers

Like tomatoes, peppers are technically a fruit but are so tightly linked to vegetable-based dishes and crops that many people think of them as vegetables. Peppers share many similarities to tomatoes in their growing preferences. Peppers like a pH level between 5.5 and 6.0 and a temperature in the range of warm to hot.

You can start peppers from seed or seedling. It takes two to three months for your peppers to mature. When considering what kind of peppers to grow, know that jalapeno, habanero, mazurka, fellini, nairobi, and cubico peppers all do fantastic in hydroponic growing.

Radishes

Like lettuce, radishes are one of the easiest plants to grow, whether it is in a traditional soil garden or in a hydroponic setup. As suggested in the last chapter, radishes are best grown from seed rather than seedling and it only takes between three to seven days to begin seeing seedlings from them. Radishes grow well in a setup with lettuce because both plants like cool temperatures and a pH level between 6.0 and 7.0.

What's really good about radishes is that they don't need any lights, unlike most plants. This means that if the cost of getting extra lights is too much for you right out of the gate, radishes offer a way of trying out hydroponic gardening before dropping the cash. What's craziest of all is that radishes can grow super-fast, sometimes being ready to harvest within a month!

Spinach

Another plant that grows well in combination with lettuce and radishes is spinach. Spinach enjoys cool temperatures and a pH level between 6.0 and 7.0, so it fits in perfectly. It needs a little more light than radishes do but it doesn't require very much at all.

It'll take about seven to ten days to go from seed to seedling with spinach and can be ready to harvest within six weeks. Harvesting can last up to twelve weeks depending on how you do the harvest. You can either harvest the spinach in full or you can pull off a few leaves at a time. This makes spinach another great option for those first getting into hydroponic gardening.

Tomatoes

Okay, okay, we all know that tomatoes are technically a fruit. But we're looking at it here because together with the other vegetables in this section, add tomatoes and you have one great salad! Tomatoes will grow best in a hot environment and you will want to set up a trellis in your grow tray. They also like a pH level between 5.5 and 6.5.

Tomatoes come in a variety of types; from the traditional ones we're looking at here to those small cherry tomatoes that make delicious snacks. Germination can be expected between five and ten days and it will take a month or two before you begin to see fruit. You can expect it to take between fifty and a hundred days to be ready for harvesting and you will be able to tell by the size and color of the tomatoes.

Fruits

Nothing tastes sweeter than fruit that you have grown yourself. Hydroponic gardening offers a great way to grow some fruit inside the comfort of your own house. Like vegetables, there are many options available to us but we'll be focusing on those that grow the best.

Blueberries

Great for snacks, baking, and even adding vitamins to your morning meal, blueberries are a fantastic crop to grow. However, blueberries can be quite difficult to germinate from seeds so it is recommended that you transplant blueberry plants instead. Blueberries are one of the slower plants to begin bearing fruit and can even take more than a year to get to the point of producing. They like a pH level between 4.5 and 6.0 in a warm climate.

Strawberries

The most popular of all the fruits that we can grow hydroponically, you can find strawberries grown in smaller personal hydroponic setups and in the larger commercial growing operations. Preferring a warm temperature and a pH level of 6.0, strawberries grow best in a nutrient film technique system.

Strawberries that are grown from seeds can take up to three years to mature to harvesting levels, meaning that, like blueberries, they are a long-term crop. Together, blueberries and strawberries make great fruit crops, which can produce for several years if you are able to give them the growing time they need.

Herbs

Herbs make a great addition to any hydroponic setup. This is because it has been shown that herbs grown hydroponically have twenty to forty percent more aromatic oils than herbs that have been grown in a traditional soil garden. This means that you get more out of your hydroponic herbs. This allows you to use less for the same end goal in your cooking, which means that your herbs will last you longer.

The best system for growing herbs is the ebb and flow system. Hydroponic herb gardens are becoming the norm across the world because of their effectiveness. There are now even restaurants that grow their own hydroponic herb gardens on site because it is the most effective way to get fresh herbs of amazing quality.

Hydroponic Nutrition

In this section, we'll turn our attention towards the nutrient solution that we use to fill up our reservoirs and provide our plants with what they need to continue to grow and stay strong. In order to get an understanding of this important component of our hydroponic systems, we will explore macro and micronutrients, the importance of researching the needs of our plants, and how we go about mixing our own solution so that pH levels and electrical conductivity are in proper ratios.

What is a Nutrient Solution?

When we talk about the nutrient solution we use in our reservoirs, we are speaking about a properly proportioned liquid fertilizer. While there are a lot of commercial options available on the market today, we will be exploring how we go about mixing our own. This way, even if we decide to go with a store-bought option, we know how we can get the most control over our hydroponic garden's nutrition.

When it comes to growing plants, there are sixteen elements that combine together from the nutrients we use, our water, and the oxygen in the air. A nutrient solution replaces those nutrients that would be found in the soil by combining them together in our water.

Primary Macronutrients

When we speak about primary macronutrients, we are referring to those nutrients that our plants require in large quantities. For humans, macronutrients are fat, protein, and carbohydrates. While plants do care about these components, it is more for how they produce and handle them inside of themselves. When it comes to the nutrients they are after, plants love nitrogen, phosphorus, and potassium. We want to make sure that we have proper ratios of these big three so that our plants can stay at their healthiest, produce bigger yields, and continue to grow.

Nitrogen

Found in amino acids, chlorophyll, and nucleic acids, nitrogen is an element made up of enzymes and proteins. While humans like protein in its pure form, plants like it when they get it through nitrogen. If your plants aren't getting enough nitrogen then they will have a lower protein content. Too much nitrogen, on the other hand, leads to darker leaves and it adds to vegetative plant augmentation.

We want to make sure that our plants have a proper nitrogen balance because this will make sure that our plants are stronger, make better use of their own carbohydrates, stay healthier, and manufacture more protein.

Phosphorus

Phosphorus is actually a major element in the RNA, DNA, and ATP system of our plants. This is a lot of scientific jargon to say that phosphorus is super important to our plants. A deficiency of phosphorus can cause our plants to take longer to mature. Not only that, poor plant growth and root growth can also lead to a reduced yield and see the plant's fruits drop off before they are mature. Likewise, too much phosphorus can lead to a lack of zinc (a micronutrient) in our plants.

Our plants want to get enough phosphorus so that they can better make use of photosynthesis. It also helps in controlling cell division and in regulating how they make use of starches and sugars.

Potassium

The last of our three big macronutrients, potassium is slightly less important than nitrogen and phosphorus. This should not be taken as an excuse to ignore the potassium levels in our nutrient solutions. When our plants don't get enough potassium, they are at risk of having weaker stems and a reduced yield. Likewise, when we have too much potassium, we mess with the magnesium uptake of our plants.

When our plants get the right amount of potassium, they are using the water from our reservoirs to the best of their ability. Potassium also helps with our plants' resistance to disease, how they metabolize their nutrients, and how they regulate excess water.

Micronutrients

When we speak about micronutrients, we are referring primarily to seven different nutrients that our plants like to have. These are boron, chlorine, copper, iron, manganese, molybdenum, and zinc. Together, these micronutrients aren't nearly as important as the macronutrients but they are still very important.

Typically, horticulturalists only add micronutrients when their plants show signs of some sort of deficiency. However, before you start adding micronutrients into your mixture, you want to make sure that the issue is actually with the nutrients themselves. For example, a possible deficiency can also be caused by pests or poor pH levels. If we start adding micronutrients into our mixtures when the problem has nothing to do with the micronutrients, then we are risking damage to our plants. For this reason, you should first consider all the possible causes and rule out as many as you can before you start reaching for micronutrients.

Mixing Your Own Solution

The first thing we need to do when mixing our own solution is to figure out exactly what our plants need. We saw how we did this above in the section titled "What is a Nutrient Solution?"

The information that you found in this section will let you know exactly what your plants want. We will take that information and use it here to fill out the specifics of this approach.

Chapter 7: Organic Farming

When you plant and care for a garden, you experience something that is truly gratifying and fulfilling. As you watch flowers bloom, herbs grow, and both fruits and vegetables ripen, you are satisfied with and proud of your accomplishments. Not only are the flowers beautiful and calming, but you are able to use the produce and herbs you grow to make more delicious meals, sustain your family, save money, and benefit the environment. By gardening, you get to experience something that, along with being helpful and fulfilling, is also peaceful and fun. This is only compounded upon when you choose the organic method.

By choosing organic and natural options to garden with, rather than the synthetic chemical-based fungicides, herbicides, pesticides, and fertilizers, you can benefit the earth and environment. It may seem like a small change, but as people continue to make the switch to organic, the beneficial effects will grow and the earth will gradually heal from the chemicals we have poisoned it with. By choosing organic gardening, you can naturally fertilize the earth with options that will refuel it with the minerals and vitamins it requires in order to grow healthy plants that will nourish those who eat them.

It doesn't matter if you have a green thumb and years of experience with gardening or if you are a complete beginner. Maybe you tried to grow some plants in the past and they always died–that's okay! With the right approach and knowledge, anyone can grow a successful garden yielding fruits, vegetables, herbs, and flowers. All you need are the right tools of knowledge, which are provided within these pages. Let's explore some of the basics of the organic method to get you on your way to success!

Plan Your Garden

Before you focus on the ever-important soil, watering, and planting of your garden you first must plan. When gardening, it is important to know the size of your garden, the hours of sun access, wind, and more. All of these factors will greatly affect what you can plant and how well your crops grow. Thankfully, with a little knowledge, you can plan the perfect garden.

1. Garden Size

    If you live in an apartment then you really only have one option for a garden, which is container gardening. Thankfully, there are many options for container gardening and you would be surprised by the crop size you can yield.

If you live in the suburbs or the city and have an average-sized yard then you should find that your garden is easily accessible from your house. Even if your backyard is on the smaller side, you should be able to plant a decently large garden if you practice gardening by the foot. With this gardening method, you can maximize your gardening space and crop yield with a little extra planning.

Lastly, if you live on a large property you can have an especially large garden. You will likely be able to not only provide enough crop yield for your family, but even excess to sell at the local farmers' market or to gift to neighbors if you desire. You want to be able to examine the garden on a daily basis so that you know when it needs to be watered, weeded, or harvested.

2. Sun Exposure

   There are some plants that need shade, although, most crops require ample sun exposure. Unless you are planning on growing shade-hardy plants, you need to ensure that the gardening location you choose receives a minimum of six hours of full sunlight on a daily basis.

3. Wind Exposure

   Heavy winds can greatly damage a garden by breaking the plants and carrying away the precious and nutritious topsoil of the garden bed. Therefore, try to find a gardening location that is not overly windy. If you are unable to avoid planting in a wind-resistant location or live in an area that gets high winds, then you may need to build a windbreak for your garden. There are many types of windbreaks, which we will detail further on in this book.

4. Find Your Hardiness Zone

   When planting, it is best to choose plants native to your area. These plants are non-invasive species that are designed to grow well where you live. You can easily find these plants by looking up the United States Department of Agriculture's (USDA) local hardiness zone map. This map is organized by ten different numbered zones. Once you know the number for your zone you can easily find which plants grow well in your local area based on temperatures and other factors.

5. Create a Garden Wish-List

    After you know your hardiness zone, create a list of all the plants within your zone that you hope to grow. This includes vegetables, fruits, herbs, and flowers. Some plants are annuals (grow year after year) while others are perennials (die-off at the end of each season). You may want to organize your list between these two categories so that you can more easily decide where to plant everything.

6. Create a Garden Map

    Measure out the exact garden space you will be using and then create a to-scale map on graph paper. When creating this map, you want to create a layout of where you will plant everything, keeping in mind to arrange it in a symbiotic manner. While doing this, you will need to know exactly how much space each plant will require, which you can usually find in organic seed catalogs and on seed packets.

7. Schedule Your Planting

    The seed catalog you purchase from should have details on when your seeds should be sprouted and transplanted. This will be impacted by your local frost-free date. You must ensure that you plant after this date, otherwise, the seedlings will die. In order to find your frost-free date, you can easily search online or call your local extension office. DavesGarden.com has a handy frost-free date page where you simply have to type in your zip code to find when your standard frost-free date occurs.

Soil

The most important thing when practicing organic gardening is to start with your soil. In chemical-based gardening, people rarely pay attention to their soil. They just cover it up in chemical-based fertilizers and pesticides. This leaches the nutrients out of the soil and causes damage for years down the road. But, organic gardening literally starts from the ground up. If the soil is healthy and nutrient-dense then you are more likely to have healthy and thriving plants.

1. Ensure You Have Enough Soil

    Many people are surprised to find that the soil in their yards is not ideal for gardening. Many yards, especially within the previous fifty years, do not have enough topsoil to grow plants well. This is because construction companies will often build the house and then only add a thin layer of topsoil afterward in order for grass to grow.

    To create a layer of topsoil, simply combine ten percent peat moss, thirty percent topsoil, and sixty percent compost. Some people may confuse compost with manure, but they are different products and not interchangeable. You can combine this topsoil mixture in the ground, in a container garden, or in a raised garden bed.

2. Find Your Soil Type

    There are multiple types of soil including sandy, loam, and clay. To determine which type of soil you have, pick up a small handful of soil that is moist but not wet. Firmly squeeze the soil in the palm of your hand. If the soil falls apart as soon as you open your palm, then it is sandy. If the soil holds its shape but crumbles upon being poked, then it is loam. Lastly, if the soil holds its shape both when you open your hand and when poked then it is clay.

3. Learn the Soil's pH and Nutrients

    It is important to know your soil's pH, because if your soil is overly acidic or alkaline then crops will have trouble growing. The pH scale runs between zero and fourteen with a reading below seven being acidic, seven being neutral, and above seven being alkaline.

4. Nourish Your Soil

    After you get the test results of your soil back then you can use the information to improve your soil. To do this, read

Chapter 3: Healthy and Rich Soil to best know how to adjust the nutrients and pH level of your soil.

5. Enhance Your Soil

After you fully nourish your soil with nutrients and balance its pH to an optimal level, then you want to enhance the soil to further improve growth. You can do this by adding in compost and manure. When adding these to your garden bed, you want to mix it well into the top 8 - 10 inches of soil.

Compost is full of nutrients, but also has a neutral pH level of seven, making it optimal for all gardens. You can either produce your own compost or purchase organic compost. Any compost you use should be a rich dark brown.

Manure is also full of many nutrients. However, the type of manure you use has to be taken into account based on your soil's pH level. Depending on the type of animal that produced the manure the pH can vary. Sheep and chicken manure tends to be highly concentrated while cow and horse manure are milder. If you live on a farm and hope to use the manure from your own animals you must let it compost for at least ninety days before using, as it contains many seeds from ingested weeds that you don't want sprouting in your garden.

Chapter 8: Tips and Tricks to Grow Healthy Herbs, Fruits, Vegetables, and Mushrooms

Below are some tips that will ensure you have the optimal harvest:

Check the Water Quality
One factor that determines whether your plants can absorb nutrients is the level of mineral salts available in your water.

Hard water is known to contain high levels of minerals. This makes it difficult for nutrients to dissolve. Thus, if you use this water, you must ensure that you filter it to make it fit for use.

Another thing you should check is the water temperature. If the temperature is higher than 85°F, you need to cool it back down. You can do so using a cooling fan.

Purchase a pH Meter
Another factor that affects the way plants absorb nutrients is the pH level. In hydroponic systems, the pH should normally be between 5.8 and 6.2. If your water does not meet these standards, you can use pH up or pH down chemicals to adjust the water. Use a pH meter to test the pH of the water you are using before making any adjustments.

Engage in Pest Control
Hydroponics gardening reduces the risk of pests but it does not eliminate it. Thus, you have to thoroughly familiarize yourself with your hydroponics system to be able to see any undesirable changes. Things such as bugs are easy to spot whereas you may have to check carefully for other things such as fungi.

Prepare Your Seeds for Germination
Some people prefer using seeds instead of cuttings. If you do so, you can help your seeds grow better by spurring germination. You can do this by placing your seeds on

moisturized paper towels. Wrap them in the moist paper towel and then place them in a plastic bag. Once done, carefully place the bag in a dark place for 24 hours. This will spur germination.

Place the Growing Medium Correctly
You need to plant your seeds properly in order for them to grow. This requires you to submerge the growing medium in the nutrient solution. However, don't submerge all of it. Only submerge 70 percent of it in the water. Once done, the seedlings should sit above the waterline. The idea is to give them a little room to breathe and take in oxygen.

Select the Right Nutrients
We have seen that plants need nutrients to grow. However, did you know that there are hydroponic nutrients available for specific plants? Instead of buying general nutrients, you can actually purchase nutrients specifically designed for the kind of plants you are growing, and when it comes to fertilizer, it would be best to stay away from it if you are just beginning hydroponic gardening. However, if you insist on using fertilizer, make sure you use hydroponics gardening fertilizer.

Also, ensure you label the nutrients you are using and write down how much you used. This may come in handy when you go through another planting season.

Study Your Market
If you plan to sell your produce, you must first engage in a little bit of market research. This is especially so if you are planting perishable produce. Remember, it takes money to grow crops. It would be sad to have food go to waste because you did not do your market research. You need to know where you can sell your crops even before you start planting them.

Even if you are just planting for your own use, it would be appropriate to study various ways of storing your produce to make them last longer. For example, some plants can be sun-dried and others can be cooked and frozen for later use. If you know how to store your crops, you will have a better experience with hydroponics gardening.

Start Small
When it comes to hydroponics, a single mistake can derail your progress and cause you to have poor or no yields. As such, it is better to familiarize yourself with various types of hydroponics gardening before engaging in large-scale farming. This will allow you to know what to expect and learn from your mistakes even as you gear up to expand your garden.

Also, when it comes to your hydroponics system, you need to study it section by section. At times, you may think that the system is too complicated but once you study it and get to know how it works, it becomes easy to operate and manage.

Thus, don't get overwhelmed. Take the time to gain the necessary knowledge that will allow you to grow your vegetables and fruits.

Imagine harvesting nearly half of a ton of tasty, stunning vegetables from a 15 x 20 foot plot, a hundred kilos of tomatoes from 100 rectangular feet, or 20 kilos of carrots from 24 rectangular ft. Yields like these are less complicated to achieve than you might imagine. The secret to awesome-effective gardening is taking the time now to devise strategies to be able to work for your garden. Here are seven excessive-yield strategies gleaned from gardeners who have discovered how to make the most of their garden space.

Plant in Raised Beds with Rich Soil
Expert gardeners agree that building up the soil is the single most important thing in increasing yields. By way of the use of much less area for paths, you have more room to grow plant life.

Raised beds save you time, too. One farmer was able to harvest 1,900 pounds of clean veggies. That's a year's delivery of food for three people from approximately three general days of work! How do you raise beds? Plants grown close collectively crowd out competing weeds, so you spend much less time weeding. The nearby spacing additionally makes watering and harvesting more efficient.

Round Out the Soil in Your Beds
The shape of your beds can make a difference, too. Raised beds become more area-green through lightly rounding the soil to the shape of an arc. A rounded bed that is five times larger across its base, for instance, may provide a 6-foot arc above it. That base won't appear like much; however, multiply it through the duration of your bed, and you'll see that it can make a huge difference in general planting place. In a 20-foot-long bed, for instance, mounding the soil inside the middle increases your general planting region from 100 to 120 square feet. That's a 20% gain in planting area in a bed that takes up an equal amount of space. Lettuce, spinach, and other veggies are the best plants for planting on the rims of a rounded bed.

Plant Crops in Triangles Instead of Rows
As an alternative, stagger the plants by planting in triangles. Doing so, you can fit 10 to 14% more plant life in each bed. Just be careful not to put your plants in too tightly. As an example, for one researcher the harvest weight was consistent with the plant doubled. (Remember the fact that weight yield in line with the rectangular foot is more essential than the number of plants per rectangular foot.) Overly tight spacing also can strain plants, making them more at risk of sicknesses and bug infestation.

Grow Climbing Plants to Capitalize On Space
Growing greens vertically also saves time. Harvest and restoration move quicker due to the fact you can see precisely where the fruits are. Fungal diseases are also less likely to have an effect on upward-bound plants that sway to the air that circulates the foliage. Tie the growing vines to the trellis. However, don't fear securing them if heavy. Even squash and melons will broaden and get thicker stems.

Choose Compatible Pairings
Interpolating well-suited vegetation saves space, too. Bear in mind the conventional local American mixture, the "3 sisters:" corn, beans, and squash. Robust cornstalks aid the pole beans, even as squash grows freely at the floor underneath, shading out competing weeds. Other well-suited combinations encompass.

Time Your Crops Carefully
Succession planting allows you to grow more than one crop in a given area over the course of a growing season. After which, they develop better inside a single growing season. To get the most from your succession plantings:

Use Transplants

A transplant is already a month or so old when you plant it, and matures a lot faster than a seed sown without delay in the garden.

Pick Fast-maturing Types
Fill the soil with a ¼ - ½-inch layer of compost (about 2 cubic ft. for 100 rectangular feet) whenever you replant. Work it into a few inches of soil.

Stretch your season by covering the beds. Some additional tips include:
- Recognize what tools you need and why.
- Recognize the nutritious requirements of your plants.
- Recognize the light/photoperiod requirements of your flora.
- Use an expert 3-part hydroponic vitamins product .
- Do not use additional nutrient additives your first time.
- Have a written plan/feeding timetable before you start.
- Have all the vital tools and vitamins before you begin.
- Grow indoors when it's 55°F or colder outside.
- Keep the ballast for your lights in a different room.
- Take a look at and regulate your nutrient reservoir solution every day.
- Minimize light exposure on your nutrient solution.
- Have an additional reservoir of pure water waiting for your next nutrient trade.
- Exchange your water and vitamins completely every two weeks.
- Use a digital timer to control your daylight vs. dark period.

- Keep your dark period completely dark and uninterrupted.
- Clean and sterilize your system between plantings.
- Quarantine new vegetation for two weeks before adding them to your garden.
- Do not go into your garden after touring another garden or being away.
- Do not permit pets in your garden.
- Go into your garden only after a bath and a fresh change of clothes.
- Make any site visitors to your garden comply with these same rules.
- Place a screen or filter over your air consumption and exhaust equipment (if outside).

Chapter 9: Differences between Hydroponic and Aquaponic Methods

Aquaponics

Aquaponics is a symbiotic system for cultivating plants and fish for the good of all interested in the safe, natural, yet unspoiled climate and soilless environment. Not only will it make your table healthy with delicious fruit and vegetables, but you can also choose to harvest your own fish, an outstanding and nutritious source of protein that will complement your healthy plants diet. Whether you prefer fish as pets or food or a bit of both, it is up to you.

The aquaponics theory is straightforward: grow plants, fruit, and vegetables under hydroponic conditions without pesticides by growing fish instead of providing the plants with nutrients. While modern practices derive mostly from research conducted in the 1960s, the word "aquaponics" is probably as old as farming itself, which may have evolved as a result of observation of symbiotic systems in nature. Whatever its roots, industrialized farming has been practiced for thousands of years in Southeast Asian rice paddies. Fish in rice fields help nutrients for the crops to grow, and the plants help keep the water clean for the fish.

The ancient Aztecs of Mexico also developed a technique for planting crops on floating rafts in near soil conditions, and have lived alongside Lake Techotitlan where they needed arable land. Nutrient-rich soil was dredged from the bottom of the lake and spread over rafts where plants grew. As the seed ripened into plants, its roots would penetrate the soil and the shoals into the lake underneath in which the fish are abundant. Essentially, the Aztecs used natural resources on hand to practice large-scale aquaponics.

Hydroponics Modern hydroponics can be traced back to Europe in the 17th century, but much earlier origins have been argued.

The example of the Aztecs is frequently quoted as an early form, and it is a good example that the Aztecs practiced nearly soilless agriculture by drawing nutrients directly from the water. In this case, the hydroponic portion was the plant subsystem. Other, older examples, frequently mentioned, are Babylon's "hanging" gardens, which were supposed to be fed by water from the river below the Euphrates. The hypothesis is that some systems were used to move the water to the top of the gardens, and the whole system was watered with a waterfall or trickle to supply each plant in the chain. However, according to ancient writers, the system was not actually smooth or "hanging" (Diodorus Siculus, Strabo, Quinto Curtius Rufus, Philosopher of Byzantium). There may have been less soil than in natural circumstances, but the descriptions make the terraces more similar to giant grazing plants over suspended plants than actual hydroponic systems.

Ancient Egyptian hieroglyphs seem to describe the process of plant cultivation in water, but it appears that there is not much information about what exactly this process was. Sometimes the Roman Emperor Tiberius, who used hydroponic techniques to grow cucumbers in proto-green houses, was credited in the first century. This might be an argument for early greenhouses. Finally, Marco Polo, an adventurer and trader of the 13th century, had returned from his journey to the Far East, believing that plants in China were developing on floating beds. It is doubtful, but it is uncertain whether it was an example of 'true' hydroponics or some sort of aquaponics technology similar to that of the Aztecs. As we saw, Chinese and other far eastern cultures understood organized farming and gardening well.

All of this is told, and despite the little evidence, it would indeed be very surprising if ancient cultures had not experimented with growing plants in the water and on the water. Examples of unhealthy, naturally floating plants are found in nature, so why did they not try it? The best evidence, however, appears to show that more successful models involving aquaponics or integrated farming principles were more than likely.

Why choose Hydroponics for Aquaponics?

Aquaponics and hydroponics are often represented as competitive models: you have to choose one, but that's a false dichotomy. In fact, aquaponics is a hydroponic farm that can be traced to ancient civilizations for hundreds, including thousands of years. It is a natural and non-chemical type of hydroponics.

Even the words used to describe them are the same: hydro is the Greek water form, aqua is its Latin version, both end with "ponic," which comes from the Greek ponos for labor or work. Nonetheless, both terms were recently created and have different histories: the current term "aquaponics" comes from a combination of aquaculture (fish farming) and hydroponics. Aquaponics is just hydroponics. It gives the impression that the aquaponics is more recent and, in some respects, derived from the hydroponics, but as we've seen, it indicates that water pumps or integrated agriculture are actually ancestral to both modern versions. This is true of the ancient practice.

Modern hydroponics removes the natural symbiotic element from the equation and replaces it with a chemical solution for the most part. This could be necessary and feasible for space exploration or in other situations where a symbiotic system is not feasible, but aquaponics is the superior choice, in my opinion.

Both processes in soilless or near-soilless plantings are highly successful, so why choose aquaponics? Well, the absence of chemicals enhances and improves the taste of plants grown in aquaponics, plus if you have chemical sensitivity, hydroponics cannot be for you anyway. However, fish can be used as a secondary food source if you are inclined to farm them as well. Ultimately, you grow two excellent food sources for the price of one in taking care of the fish for a few minutes a day. If the fish are in good shape, the plants require little or no attention. Talk about efficiency! Some people attach themselves to fish or to vegetarians, and that's great, you're still in perfect symbiosis: feeding fish, feeding plants and filtering fish water, in an infinite, self-sustaining cycle.

You can start with a few plastic containers in your backyard or garage or convert the project into a large commercial enterprise and still have experience with a natural, hydroponic system. In no time a simple system will pay for itself. With just a few efforts to get things started, you end up with a near-self-supporting system that produces organic products of high quality without the high price of organic products. The joy you get from growing your own food is also a nice bonus. Involve the entire family and have fun. The components are relatively inexpensive, and you can avoid the cost of a kit or hiring professionals if you build it yourself. If you're interested in gardening or would like more options for your daily diet, aquaponics is an option that you might want to take seriously.

Chapter 10: Types of Aquaponic Systems

Most of the time, you will only see one type of aquaponics system in people's DIY setups, and that is the media filled bed. There are a few other systems we are going to look at here. We will cover the most popular one first for the backyard gardener.

The Media Filled Bed

This is generally considered the easiest form of aquaponics and the best one for beginners to start with. Why? Because it acts as grow media and a bio-filter all in one. The media filled bed is also called a grow bed.

A grow bed is mainly used as a place to house the nitrifying bacteria. The bed will also be used as a growing/securing media for the plants.

Generally, grow beds are used for larger plants that have bigger roots to keep them upright. Because the recommended depth of these beds is 12 inches deep, they are perfect for growing tomato plants or starting fruiting trees.

There are two types of systems for the media bed. The first one is called continuous flow, while the other is known as flood and drain or ebb and flow.

Continuous Flow

This system will keep your grow beds filled at a constant level. This means the media will be submerged all the time without allowing oxygen to access the roots. This could be problematic if you don't have enough aeration in your system. The bacteria need oxygen to thrive. For this reason, I don't recommend using a continuous flow system. It doesn't guarantee seamless operation, and I would be checking the oxygen levels in the water all the time.

Flood and Drain

Instead, I suggest using the flood and drain method as it gives peace of mind, and you don't have to monitor oxygen levels very often.

The flood and drain cycle uses a siphon method to allow the entire media bed to fill with water and then drain again using a siphon. This way, the growing media, and the roots, will be exposed to the air in every cycle. This promotes bacteria to grow and the roots to be exposed to oxygen.

The Siphon

So how is the water drained, you might ask?

We use something called a siphon. There are two main types of siphons. I'm going to explain the easy version first and then the more popular but complex bell siphon.

Option 1 - Timer

With this siphon, you need a timer for your pump. This way, your system requires less energy while still doing the same function as the bell siphon. This method is preferred for beginners.

In this image, you can see the water from the fish tank being pumped up to the grow beds. Here, the water level is at its highest. The excess water will be drained by the 'open pipe for overflow.'

There are little slots drilled or cut in the middle of the big standing pipe. This allows the water to flow there. The slots need to be small enough to not let grow media (rocks) get through.

There is a smaller standing pipe inside the big standing pipe that is directly connected to the fish tank with a bulkhead fitting or unseal. This pipe will allow the water to flow back to the fish tank and act as an overflow. Remember that the diameter of the small standing pipe must be big enough to function as an overflow (at least one inch).

The timer initiates and the pump moves the water from the fish tank up to the grow bed. The grow bed starts to fill with water.

The water level in the grow bed rises to the level of the overflow (the small standing pipe). The excess water that gets pumped in the grow bed will exit the overflow back to the fish tank. Allow it to overflow for a few minutes.

The timer will stop, and the pump will shut down. The water will slowly drain through the small hole at the bottom of the small standing pipe. After the grow bed has fully drained, the timer initiates again, and the grow bed will start to fill up, repeating the cycle.

Because every grow bed is different in size, you need to watch how long it takes for your grow bed to drain and fill up fully. Then, adjust your timer.

While the bed is filling up, the water will drain from the small hole in the standing pipe. However, the hole at the bottom of the standing pipe will be quite small. You need to make sure that the flow rate of the small hole is less than the flow rate of the water that is being pumped into your grow bed.

The size of this hole is crucial. Drill it too small, and your grow bed will drain slowly. Drill it too big and the grow bed will fill up very slowly.

Option 2: Bell Siphon

The bell siphon is a great way to manage a flood and drain cycle naturally. We've already looked at its basic function: to help flood and drain the grow bed. But let's look at it in a little more detail.

It is always a good idea to have an overflow pipe feeding back to the tank. This will prevent the water from overflowing your grow beds if you have a problem with the drainage through the bell siphon.

The basic bell siphon will have a ¾-inch (19 mm) pipe running up to approximately one inch below the top of the growing media. You will then have a bigger inverted pipe (dome) over the pipe, which sticks just out of the water. There needs to be a gap at the bottom of the inverted pipe to ensure the water can get to the drain pipe.

The drain pipe you have just installed will be full of air. As the water moves up in your tank, it will eventually reach the top of the pipe. This will block the air and cause a vacuum in the pipe. The vacuum will then suck the water down the pipe and keep moving water until the level is below the top of the pipe. At this point, the pipe draws in air instead of water and stops draining the water from your grow bed to your fish tank. This is what we call: 'the siphon breaks.'

After the siphon breaks, the grow bed will fill up again.

There are two key points that need to be addressed: getting the siphon to start with a lower water flow and giving it a good flow rate.

It is possible to reduce the siphon starting point by adding a pinch in your exit pipe. However, this will also reduce the flow rate of the water, which is not desirable.

Instead of a pinch, you could add a 90-degree elbow into your system just below the tank. This would create a turbulence spot, which would improve the efficiency of air movement and reduce the amount of water needed to start the siphon. However, this will also decrease the equilibrium flow and not aid your ability to create a reliable siphon effect.

The best solution is to increase the size of the pipe for the last inch or two as it reaches the top of your grow bed (use a ¾" to 1½" adapter). This simple adjustment will turn the main pipe into a pinch effect without reducing the flow rate. In short, you'll get low flow in the beginning, which increases in power once the siphon has initiated.

The speed of water removal will ensure the siphon breaks and the flood and drain part of your system works efficiently.

This simple technique works as the larger area at the top of the siphon pipe creates a strong starting flow combined with the narrowing width of the pipe and, if possible, a longer pipe outside of your tank. This will reduce the amount of water you need to start your siphon effect.

The equilibrium flow will be stopped by the continual residual flow and the funnel at the top of your pipe, causing a water imbalance that will shut down the siphoning effect.

Tips to Get Your System Draining Properly

Now that you know how to make your own siphon, the following tips will help you to ensure it is working properly.

Expanded Shale

This is crushed before it is heated to high temperatures. The effect is the same as with the clay pebbles. It is clean, porous, and pH neutral. Expanded shale is also very light and has a large surface area, making it very effective to house bacteria.

Lava Rock

Lava rock is surprisingly light and extremely porous. It has been used as a growing medium for many years. The nature of this rock ensures excellent surface area. They are also pH neutral. Lava rock is frequently used in garden ponds to house bacteria, making it easy to source them from a local pond store. They generally have very sharp edges, which means they can cut your hands and even damage the roots of plants if you move the rocks around once they are in place.

River Rock

River rock is a very good and inexpensive alternative to expanded clay. The only downside to it is its weight. If you are on a budget, river rock is the perfect choice for you. It's easy on the hands and available literally everywhere. I recommend using ¾-inch river rock.

Gravel

This is a great option if you are on a tight budget, as it tends to be the cheapest and easiest grow media you can get. However, it can give you a few problems.

Gravel tends to bunch together exceptionally well. This makes it a dense material for roots to find their way through. While this shouldn't be an issue if you have tall plants with long roots, it will be a problem for most of the smaller plants. If you use gravel, you should always wash them first and do a vinegar test before putting it in your tank. It is possible that there will be limestone in it, which will increase the pH of the water.

How to do a Vinegar Test

Take a handful of washed gravel and put it in a cup (preferably transparent). Fill the cup with white vinegar. If you see fizzing (like soda), there is limestone in it. If you do see signs of limestone, you shouldn't use the gravel. Test a few batches of it before deciding. This may seem like a lot to take in, but once you understand the various growing media and the different setups, you can start deciding which ones will suit you best. Most people use expanded clay for their system because of its weight and the fact that it is easy to handle.

Chapter 11: How to Build an Aquaponics System – Step-by-Step Guide

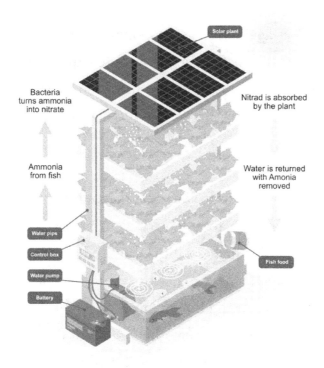

By now, you should be getting excited! You have been enlightened as to what an aquaponic system is and how it can benefit the environment as well as creating an organic food source. You should also have an idea of all the basic equipment you need to get started on building your own system.

Of course, when you're first setting up, you're probably not thinking about creating enough food to feed the whole neighborhood. Succeeding in growing plants and keeping your fish healthy is a good starting point.

For this reason, you may want to start very small. However, I would caution against going too small; you'll simply create more work for yourself when you want to expand your setup.

That's why it is good to follow this step-by-step guide. I recommend starting with a simple IBC setup.

Step 1 – Size

There are many reasons to start an aquaponics setup. You may need to grow enough food to feed your family, or you may want to start selling high-quality local produce at the market. The point is that you have to decide what size yield you would like to achieve. Knowing how many plants you want to grow will allow you to work out the other details of your setup.

Your plants should be spaced according to the size you expect them to grow. For example, a head of lettuce will need approximately 8 to 12 inches of growing space. The primary concern with plant spacing is to know which plants you are growing and how much room they need when they are ready for harvest. Here is a general guideline that shows how many plants you can get per square foot:

- Basil – 4
- Carrots – 8
- Celery – 1
- Tomato – 1 per 4 square feet
- Lettuce (leaf) – 4
- Lettuce (head) – 1
- Onions – 4
- Spinach – 4

Be warned that herbs like mint will spread quickly. It's best to keep an eye on these. Otherwise, they will take over your grow bed. If you are growing the plants from seed, there will be a recommended spacing recommendation on the package of seeds. Once you've established the number of plants you want and calculated the space they need, you'll have an idea regarding the size of the area you need to grow them (the growing pace).

Next, you need to decide what kind of system you are going to use. I recommend choosing grow beds for your first system.

Step 2 – Growing Area

Once you've calculated how many fish you can have, you need to decide which type of system you are going to use.

If you're doing this for the first time, it's a good idea to start with a standard IBC grow bed. However, if you're feeling adventurous, you can opt for the deep-water culture that was discussed earlier in the book.

The standard media bed can be created from:
- IBC containers
- 55-gallon drums cut lengthwise (barrel phonics)
- Wood and liner
- Concrete and liner
- Plastic containers
- Bathtubs
- Plywood/cinderblocks and epoxy paint

The DWC system is usually created with wood and pond liner or concrete and food-grade epoxy paint. It should be at least 12 inches deep with air stones on the bottom. The width will depend on the material you will use as the rafts. The recommended material is medium to high-density Styrofoam that's not fire-retardant (because of chemicals) and 2 inches thick (50mm). You can paint it white, so it reflects the sun. This way, it will not heat up your water. I talk about using paint in aquaponics later in the book.

Step 3 – Location

Now you need to decide where you are going to grow. Part of this decision will be made by the size of your system. If it's small, you may want to set it up inside your home. If you do this, you'll need to make sure there is enough light for the plants to grow and flourish. This requires more electricity than growing outside, but because of the warm temperatures inside, you may be able to grow faster.

Larger setups will need to be outside to use the sun as a source of energy. Using lights will drive up the cost of production and will cut into your profits.

If your local weather is not consistent or you're likely to experience drops in temperature, it will damage your plants. You may need to consider building a greenhouse so you can grow year-round.

Step 4 – Connecting It All Together

Now you need to connect the grow bed and fish tank together. You'll need pipes to feed the water from the fish tank to the grow bed and another to take it back. It is also a good idea to have an overflow pipe set approximately an inch below the top of the growing media. This will help prevent an overflow if the siphon fails or becomes blocked.

If you're opting for a small system, you can place the grow bed on top of the tank and use gravity to allow the water to flow back into the fish tank.

You will need to use an aerator to add oxygen to the water, and a pump needs to be added to the fish tank.

Step 5 – Adding Water

You are now ready to add water to your tank!

Adding some water or grow media from an already established aquaponics system or pond is the best method because it already contains some bacteria.

The next option is to use plain tap water. Keep in mind that it will take longer for bacteria to populate the system when using tap water.

With tap water, there will be chlorine in it. Turning on your aerator will help to get rid of the chlorine, but it can still take several days or even weeks if you have a larger system.

If your water has chloramine, you need to use a reverse osmosis filter or a carbon filter. Chloramine can't be aired out like chlorine.

If you are using well water, you need to be wary of water hardness and alkalinity. You need to test the water and keep these under a certain level. Hardiness of the water will mess with the nutrient availability to the plants. I talk more about hardiness and alkalinity later in the book.

Step 6 – Cycling

To get the desired bacteria, you need to have ammonia in your tank. To create ammonia in your tank, we do something called cycling. There are two types of cycling: cycling with fish (safest) and cycling without fish (fastest).

Cycling with Fish

It is possible to cycle by introducing fish and letting them prepare the water. This will take approximately 8-12 weeks, depending on your system.

The fish are suppliers of toxic ammonia. If you introduce the fish first, you'll need to monitor the ammonia almost every day because there will be no bacteria at the beginning to convert ammonia to nitrites and nitrates. If the ammonia level rises above the allowable value (see next table), then you need to reduce the feed you give to the fish or even do a partial water exchange as an emergency fix.

Most of the time, you will start the system with small fish. Using small fish is never a problem when you are cycling a new system. However, if you use fully grown fish to start the system, you might run into some deadly ammonia and nitrite spikes before the right number of bacteria are present. Remember that ammonia levels are at their highest right after feeding.

Step 7 – Testing

Before you introduce your fish and plants, it is important to test the water.

The levels of ammonia (ppm) change with the pH and temperature of the water. If temperature and pH increases, so does the toxicity of the ammonia. Basically, if you know the pH and the temperature of the water, you will automatically know the level of toxic ammonia.

	pH				
Temperature	6	6.4	6.8	7	7.4
39.2°F (4°C)	200	67	29	18	11
46.4°F (8°C)	100	50	20	13	8
53.6°F (12°C)	100	40	14	9.5	5.9
60.8°F (16°C)	67	29	11	6.9	4.4
68°F (20°C)	50	20	8	5.1	3.2
75.2°F (24°C)	40	15	6.1	3.9	2.4
82.4°F (28°C)	29	12	4.7	2.9	1.8
89.6°F (32°C)	22	8.7	3.5	2.2	1.4

	pH				
Temperature	7.6	7.8	8	8.2	8.4
39.2°F (4°C)	7.1	2.8	1.8	1.1	0.68
46.4°F (8°C)	5.1	2	1.3	0.83	0.5

53.6°F (12°C)	3.7	1.5	0.95	0.61	0.36
60.8°F (16°C)	2.7	1.1	0.71	0.45	0.27
68°F (20°C)	2.1	0.83	0.53	0.34	0.21
75.2°F (24°C)	1.5	0.63	0.4	0.26	0.16
82.4°F (28°C)	1.2	0.48	0.31	0.2	0.12
89.6°F (32°C)	0.89	0.37	0.24	0.16	0.1

Having tilapia in a system at 82°F and the pH is 7.6, your maximum ammonia level will be 1.2 ppm.

When you get a reading from the pH tester, you will see the combined numbers of ammonia (NH_3) and ammonium (NH_4). Ammonia is very toxic to fish, while ammonium is not as toxic. This combined number is referred to as Total Ammonia Nitrogen (TAN).

Understanding the difference between the two is crucial to getting to know your total amount of toxic ammonia.

Step 8 – Plants

If you add plants while cycling, you will notice that they don't grow very fast. This is normal in the beginning because your system is still populating with nitrifying bacteria.

When you add plants, you need to make sure they are as clean as possible. If you want to use plants that were in soil before, you should visually check each plant to ensure it is disease and bug-free.

Step 9 – Fish

Cycling without Fish

When the nitrates start to appear, and the pH level is under 8 (preferably lower), you are ready to add the fish. Of course, the ammonia should be below the advised level (see previous table), and the nitrites should be under 0.5 ppm.

Adding fish to the tank

First, keep the fish in the bag and place the bag in the fish tank. The water temperature will slowly equalize between the two. This way, the fish are not shocked when they enter the fish tank.

After a while, open the bag and exchange some water from the fish tank with the water in the bag where the fish are. This will equalize the pH, just as the temperature did. After an hour, you can introduce the fish to your fish tank.

Don't forget that the weight of the fish is calculated on their mature weight even if you add them as fingerlings.

Feeding

Your system is now up and running! You will probably be anxiously looking at it every day. However, even though the plants grow fast, you're not going to be able to see them growing immediately.

You will need to make sure they have all the food they need. For this, you need to feed your fish; they will produce the ammonia that the bacteria can turn into nitrates and other nutrients for the plants.

We have already calculated the amount of feed we need to give to our fish in relation to the available growing area. Because smaller fish eat less, I have made some calculations to account for that.

To give you an idea of how much food is required, an adult tilapia fish will eat approximately 1.5-2% of its body weight per day. When they are fingerlings, they require 6-10% of their body weight.

Chapter 12: Sustaining the Seasons for the Greenhouse Garden

To prepare for winter, there are various things you need to check:

1 - Move Harvested Plants and Tools
Fruits and vegetables that you have already harvested need to move out of the greenhouse to create more room for the subsequent season's plants.

You should take pots, containers, seed trays, or any other tools that you are not currently using out for cleaning. Give them a proper scrubbing to remove all the dirt before taking them back to the greenhouse.

Once you have moved everything out of the greenhouse, clean the structure itself. If you have been planting directly in the greenhouse soil, remove it and replace it with fresh soil and new compost.

2 - Remove Rotten Plants and Weeds
Remove all rotting plants in the greenhouse. Pests and insects feed on the crops during the summer period, and they may lay eggs on the plants or on its leaves; therefore, removing these plants and leaves from the soil will help you get rid of pests.

If the fallen plant leaves are disease-free, you can deposit them in the garden trench and convert them to organic matter.

Remove all established weeds from your structure. You can dig them up and burn them outside. Some invasive weeds remain in the compost matter, so avoid moving the compost from one area of the garden to another.

3 - Cleaning the Inside Out
Thorough cleaning of the greenhouse is crucial. If there are still plants inside the greenhouse, move them to a warm area and scrub all corners of the structure. You can use hot water and Jeyes fluid disinfectant, which is greenhouse-friendly. Make a plastic dirt-clicking tool to enable you flick dirt from the frames.

You should clean both the inside and outside of the greenhouse, as it not only makes your greenhouse sparkle clean, but it also allows more light and warmth to enter the greenhouse when it is clean.

After cleaning, leave the door and windows open to allow fresh air in and so the greenhouse can dry out completely.

4 - Moving Tools Inside
Once the inside of the greenhouse is dry, you can return all the greenhouse tools you took out for cleaning (pots, trays, and

containers). Then, you can decide what to do with the greenhouse.

You can move frost-sensitive plants inside to give them more warmth during the winter months, especially if you have a heated greenhouse. If you don't have a heated greenhouse, you can provide more warmth to the plants by using bubble wrap.

5 - Prepare the Soil for Spring
Fall is the best time to prepare your soil while you wait for the spring season. You can do this through adding compost, manure, and rock phosphate, among other substances to the soil to boost its nutrients and texture.

You don't have to wait until spring to enrich the soil with the required nutrients. It also helps to improve the drainage system before the busy season.

After preparing the soil and making all the adjustments, you can cover the area with a plastic sheet to prevent heavy winter rains from washing away the soil amendments.

6 - Planting Cover Crops
During this period, you can sow cover crops, which helps prevent soil erosion in the area and increase organic matter in the soil bed. Planting cover crops increases nutrients in the

soil. For example, planting legumes or field peas will add to the level of nitrogen in the soil.

7 - Trim Perennial Plants

The fall season is the best for pruning perennial plants in the garden. Although this depends on the kind of plant, raspberry plants continue to grow into the winter, whereas you are better off trimming blueberry plants during spring. You can trim herbs like rosemary and thyme during the fall. Blackberries can also benefit from the winter cleaning.

8 - Regenerate Compost

After the summer harvest, you can use the compost material from the harvested plant trees and leaves to enrich the garden bed. This increases soil nutrients and can solve soil deficiencies. This practice will make your work easier as you jumpstart to the busy spring period.

Cleaning used compost in the garden makes way for new compost with more active microorganisms and green matter.

9 - Adding Mulch to the Soil

Mulching helps prevent water loss by improving soil drainage, preventing soil erosion, and preventing the growth of weeds in your garden. Winter mulching helps regulate soil temperature and retain moisture. As the weather changes, the soil transitions to match the cold weather. The ground surface freezing can affect the plant roots; therefore, adding mulch to the soil will help regulate the temperature and save the plants' roots from freezing.

If you have vegetables left in the garden during the fall season, you can add a layer of mulch to the vegetable root soil and prolong the crops' growth.

10 - Review the Growing Season

This practice requires you to review the performance of the fruits and vegetables planted during the season. You should evaluate which fruit trees did best based on the produce, and which fruit did well. From this information, you can discern the kinds of fruits and vegetables to grow in your next season.

You will also know which crops to add to the greenhouse to extend your harvest. You can choose to add crops that ripen early or late, compared to other crops you already plant.

When comparing the performance for choosing the next type of vegetables to plant, take notes on what worked for you and what didn't work and check what caused failure or success for each plant.

Chapter 13: Maintenance of Greenhouse Garden All Year Round

Top Ten Reasons to Grow a Hydroponic Garden

1. Grow indoors all year round

Hydroponic gardening is one of the best ways to produce fresh food throughout the year. It is also a great alternative for the cultivation of a number of plants in smaller areas, such as indoors. Electricity costs indoors are typically less than $3 a month, depending on the size of your device.

2. Grow more in less space

Nutrients are directly injected into the plants in a hydroponic garden so the plants can grow closer together. The roots of plants must not fight for water and nutrients. For instance, if you wanted to grow plants in the traditional way, you would space the plants in the ground to take enough water and nutrients for them to grow properly. If you wanted to grow 20 lettuce plants in the soil, the area would take around 20 square feet. In a hydroponic setup, you can grow 20 plants in an area of only 8 square feet.

3. Monitor the growth factors

All the factors, including light, water, and nutrients, can be regulated with a hydroponic system by simply turning a switch on or setting a timer. The pH of the nutrient solution can also be controlled. This is much more difficult with a conventional garden planted in the dirt. You cannot regulate soil quality, for example, without adding expensive back-fill fertilizers with potting soil. You can't control the lighting because it might be cloudy some days and sunny on others. Watering your garden is easy to control, but if you do not have water well stored, then you may need more water than you have access to or budget for, especially if you live in a dry climate. The ability to control these factors enables you to make your plants grow faster, healthier, and more bountiful.

4. No weeding

In a hydroponic system, nuisance weeds do not flourish (unless they are planted there).

5. No back-breaking work

Because there is no soil, you don't have to work the ground in a hydroponic system. Ideally, a hydroponic system is built up a few feet off the floor so that you don't get trapped in the garden on your hands and knees.

6. Freshwater scarcity

Ocean water covers more than 70 percent of the surface of the Earth, but thirsting humans rely on limited hydration and agricultural supplies of freshwater. Just 3% of all water supplies are freshwater, mainly found in the Arctic.

7. Poor soil quality

Soil is the most widely used plant medium that not only helps provide nutrients and oxygen; it also transmits water and other beneficial microorganisms to the roots. This tried and true development is still commonly used in large areas of the planet. This same soil, on the other hand, often carries risks for your plant, such as bug infestation, damaging pH, poor drainage, poor water retention, and wear due to soil erosion.

8. Hydroponics is an option for people living in an urban environment.

In general, residents of apartment buildings have no access to an open area where vegetables can be grown. Often they have a balcony or patio, but due to limited space, only a small number of plants can grow in containers. Hydroponics solves this problem by allowing more plants to be grown in a smaller space.

9. Less reliance on foreign oil

Approximately 60% of all oil imported into the U.S. is used for food production. The oil is used by farmers who grow the crops and by truckers to transport it. Our desire to eat demands considerable amounts of oil. Growing food locally through hydroponics would decrease those food miles, maintain more nutrients in the crop, help the environment, and minimize the country's dependence on foreign petroleum. It would also build a less central food chain, with a lower risk of widespread disease or terrorism. When it comes to our farms, more and more incidents have occurred nationally. For example, fresh spinach cultivated in California was exposed to E-coli bacteria contamination. When one of these food accidents occurs, people want to know more about the origin of their food. People want local fruits and produce and start thinking about how this might be possible. Seasonal consumption and purchases locally establish a stable local economy and enable everyone to eat better.

10. It's fun

Another reason to grow a hydroponic garden is that it is enjoyable and fun. If you like to tinker with things, it could be perfect for you.

15 Tips to Make Your Greenhouse More Efficient

The purpose and intention of the structure should be to optimize crop growth in the most efficient way possible if you own and operate a greenhouse. While it is true that many owners make efforts to achieve designs suited to their needs, these areas of focus will not always lead to the least amount of energy required.

1. Build a Conservation Checklist.

    Conservation is an essential component of basic operation at the cost of high energy use in today's greenhouses. Research suggests that the energy consumed by greenhouses is 75% of total energy.

2. Assess the structure as a whole.

    This is particularly important if you concentrate on climate control. Cool air or warm air can easily escape from the greenhouse. When trying to keep a certain temperature in the greenhouse, you should understand that your losses will depend on the type of cover and the age of the unit. If you want to heat the structure effectively, consider a double polyethylene cover–which can reduce your heating costs by as much as 50 percent. With a glass greenhouse, consider upgrading the structure with a double polyethylene layer–which could reduce costs up to 60% compared to the glass panels.

3. Eliminate drafts and air leaks.

It is imperative that you work to prevent any air leaks associated with the structure to ensure that your greenhouse is operating efficiently. The main place to begin is the structure door or doors. A special door closing unit should be used, or a spring mounted door to make sure the air does not enter the unit. Weather stripping should also be placed around the doors, windows, and ventilation units. Weather strip should also be placed around openings close to fans and heaters. If you find holes in the greenhouse siding or foundation, they should be repaired immediately.

4. If you want to increase the efficiency of your greenhouse, you should focus on doubling the structure coverage.
   One of the most effective and cheapest ways to do this is to bubble-wrap the interior walls of the structure. This offers what is called a "thermopane effect" that improves insulation in the greenhouse. If you have an older frame, just throw a double plastic sheet over it to reduce the infiltration and minimize heat loss by up to 50 percent.

5. Implement a conserving curtain.
   If you want an efficient greenhouse, consider a thermal curtain. Such goods will save from 20 to 50 percent in costs. If the cost of the curtain is around $2.50 on average for each square foot, the payback comes in two years.

6. Install Insulation at the Foundation

   You should take the time to insulate your foundation to reduce the efficiency of your greenhouse. The best way is to use a board made of polyurethane or polystyrene. The board should be 1 to 2 inches thick and should be placed approximately 8 inches underground to help reduce heat loss.

7. When you are interested in increasing the amount of heat stored in your greenhouse, the area behind your heating pipes should be isolated.

It is best to use aluminum-faced building paper to help radiate heat from the pipes back onto the growing area of your greenhouse.

8. Consider the location of your structure.

   To reduce your greenhouse energy consumption, put the structure in an area surrounded by trees and/or other types of structures. Over time, wind will result in a lot of heat loss in growth structures. If you put the building in a sheltered area, it is important to make sure the building always receives the correct amount of light, so that the crops continue to grow properly.

9. Place windbreaks on the north and northwest sides of the building.

   In these areas, you might put several coniferous trees or even a plastic snow fence. This reduces the amount of heat loss due to wind exposure.

10. Increase the amount of space in your Greenhouse.

    One of the most productive ways to maximize greenhouse efficiency is to increase your unit space. You can improve the amount of space you have by up to 90 percent with benches that can be moved or are peninsular shaped. You should mount stacking racks if you have small plants. Furthermore, use baskets that can be placed on rails or on overhead transport systems to grow crops.

11. Regular heating system maintenance

    If you want to save time and money by optimizing greenhouse efficiency, make sure your heating system is maintained regularly. You should make sure the boiler works optimally and is periodically cleaned. You should have a furnace regularly changed and washed. If you have a medium-sized unit, this may save hundreds of gallons of petroleum per year.

12. Use Electronic Thermostats

You should convert to an electronic model if you currently use a thermostat. In so doing, you will find that up to five hundred gallons of heating fuel can be saved each year. Using these thermostats will lead to more precise temperature measurements. Mechanical units have been measured to read sometimes up to two degrees higher than electronic devices. This could result in expenses exceeding $200.00 per year. You can avoid paying too much to control your structure's climate by switching to an electronic thermostat.

13. Install fans

    The next way to maximize the output of the greenhouse is to install fans that generate horizontal airflow.

14. Try using open-roof cooling strategies.

    If you spend a lot of money cooling your greenhouse, you can try open-roof designs. This form of design removes the need for fans and high-priced refrigeration systems.

15. Finally, if you want to optimize your greenhouse efficiency, you should install energy lighting systems. The use of moving bulbs and lights is known to be efficient devices that can save you hundreds of dollars per year.

Chapter 14: Advanced Greenhouse Garden

As previously stated, greenhouses have revolutionized modern gardening. They are extensively used for research, development, and biotechnology. The advantages of greenhouses are many and this chapter will throw some light on the various merits of using greenhouses.

Inexpensive Ways of Cultivation

Greenhouses are inexpensive ways of cultivating crops. By cultivating plants and crops in greenhouses, you don't have to wait for a certain season in which the plant grows in its natural habitat. This lets you have a constant supply of vegetables, flowers, fruits, and other types of plants all year long in some places. Planting your own vegetables, flowers, fruits, and other types of plants and crops significantly reduces expenses because it is less expensive to grow them on your own. Instead of having to go all the way to the plant's natural habitat you will able to get a better yield, regulate the temperature, and produce crops at a much cheaper rate as opposed to when you actually buy fruits, vegetables, and other plant items from the store. Greenhouses, though, are quite expensive to set up. They may be able to reduce our dependency on farmers and horticulturists and this, by itself, reduces expenses.

Consistent Temperature

Greenhouses allow for a consistent temperature. In the plant's natural habitat, there may be seasonal changes that take place. These seasonal changes include rain showers, drizzle, lack of sunlight, and other types of seasonal changes. These seasonal changes have attributed to a prolonging of the growth of the plants. Greenhouses provide a platform for an even distribution of consistent temperature. This consistent temperature can reduce the timeframe required for the plants to grow and that leads to faster-growing crops. Thus, you can harvest crops a lot faster and this will reduce your waiting time. Another advantage is that with a consistent temperature maintained in the greenhouses you are able to maximize your plant's yield. Crops that are specific to a certain climatic condition will be able to churn out for harvest in the greenhouse due to a constant temperature.

Protection of Plants
Greenhouses provide protection to plants. This is done in several ways. For one, plants are protected by a consistent temperature for them to grow in, thus leading to better yield and fewer pests. Greenhouses help to protect and reduce the intensity of the crops from harsh climates and seasonal fluctuations that can disrupt the growing of the crops. Secondly, greenhouses serve as research houses for many universities across the world; scientists are constantly working towards making the greenhouses better. With global warming, increased pollution, and many man-made hazards, several plant species have become extinct. Greenhouses allow plants that are on the verge of extinction to be cultivated. Thus, greenhouses protect plants by reducing the extent to which their existence is endangered.

Increased Yield
Growing plants and cultivating crops in greenhouses increases their yield. This is because greenhouses provide a

consistent and uniform temperature throughout the entire structure. This enables plants and vegetables to obtain the correct amount of sunlight, humidity, moisture, and other factors that are required for their growing. Thus, it can reduce the timeframe required for a plant to grow, thereby resulting in higher yield. Another advantage that attributes to better yield is that greenhouses are essentially closed glass structures. These structures are easy to maintain and the crops that are grown here can also be maintained easily because of this. This implies that there is a lesser chance of developing pests and insects that can harm the plants. Therefore, the plants are much safer. With a safer environment, they are able to increase their output.

Variety and Choices
Greenhouses enable the growth of various plants and crops that are usually seasonal in nature. Thus, there comes a lot of choice with growing plants and crops in greenhouses. For instance, a farmer can use greenhouses to grow certain types of crops that require growing only in certain conditions not naturally found on his farm. He can also grow the regular crops that are accustomed to growing in the soil on his farmland. Thus, this gives him the best of both worlds by combining the growth of seasonal crops as well as crops that are grown all year long. Similarly, in greenhouses, one can grow several types of plants. This is not limited to just the

various types of plants but the division of plants via technology.

Temperature Regulation

Greenhouses were originally used to grow seasonal plants. They basically mimic the temperatures found in the natural habitats of these seasonal plants. They were used extensively in tropical zones and areas to grow crops that are usually found in colder regions. These days, with the advancement of science and technology, greenhouses are used to grow both summer and winter crops. Today, specially designed greenhouses not only provide a consistent environment for the plants to grow, but it also accounts for regulation. The amount of sunlight, moisture, and humidity can be regulated using crafted switches and designed panels in high-end technical greenhouses.

Greenhouses can reduce the costs associated with cultivation. For instance, if a farmer produces crops in city A because the climatic conditions for growing the crops are favorable in city A, he will need to transport those crops to the city you are living in, say city B. The transport costs increase with the increase in distance. Having greenhouses in your city enables you to reduce all those costs by having the crops available nearby. Similarly, there are other costs associated with cultivation, like the storage facilities for seeds and so on. Greenhouses can serve as a storehouse for seeds for these rare and exotic plants.

Pest Free
Greenhouses provide protection to the plants by keeping them pest free. Greenhouses are specially designed to allow easy maintenance and growth of seasonal crops and plants. By cultivating in a greenhouse, you reduce the frequency of pests that can be very harmful to those seasonal crops. For instance, if you grow cucumbers in summer, then growing them in greenhouses would reduce the pests that generally feed on cucumber plants. This is because greenhouses only mimic the temperature, not the soil or the water that are additional requirements for the plants. Similarly, since these plants are grown inside a glass structure, it reduces the chance of outside species of insects and pests to get in and feed on the plants. It also reduces the intensity of grazing and eating by a variety of animals such a deer and rabbits. Greenhouses enable a pest free environment as animals cannot enter these structures and this reduces the extent of trampling over the plants.

Aesthetic Appeal
Greenhouses are used by landscape designers and outdoor designers for aesthetic appeal. Greenhouses today are not just blocks of glass structures that host seasonal plants and crops,

but they are also used for aesthetic beauty. They provide a design for your farmland or garden. They are used to beautify an existing piece of land. The lush greenery found inside the greenhouse, dotted with the colorful fruits, vegetables, or flowers, provide a good source of appeal for man's eye. These days, architects are also specially trained to build good quality greenhouses that not only are functional but also attractive. These architects build the greenhouses from unique designs and even different colors. They use various materials to make greenhouses a work of art rather than just a shelter for plants. These days, greenhouses also come in a variety of sizes and shapes to suit any interests and tastes. There are several materials like wood, plastic, glass, and various other combinations of these to specifically serve as greenhouses and act as attractively designed structures to beautify the surrounding area.

Multipurpose
Greenhouses are multifunctional and multipurpose structures. They not only allow growing of seasonal plants, but also extensive works of technology that enable research, conservation, and protection of a wide variety of plants. They also serve as storage facilities for saplings and seeds. Another great thing about greenhouses is that nearly everything can be grown in these structures. Vegetables, fruits, flowers, nuts,

herbs, and any other type of plants can be cultivated in these structures. Indoor plants and outdoor plants can be grouped together and grown here.

Controlling Humidity
If the air is humid, this implies that the air inside the greenhouse has large amounts of water in it. A higher rate of humidity slows down the evaporation rate from the plants and soil. Water is vital, especially for seasonal plants to perform regular actions like photosynthesis with adequate sunlight. Low levels of humidity can leave the plants dry and not let them perform photosynthesis and thereby the plants wilt. Greenhouses control the humidity in this way. Just like you can control the temperature, humidity too can be controlled. It provides a humid environment for the plants to avoid water loss through evaporation. Thus, the plants look a lot fresher and a lot healthier. This is especially important for plants that grow in winter seasons; they might lose a lot of water because of the direct sunlight. Thus, by controlling humidity, greenhouses help plants.

Conclusion

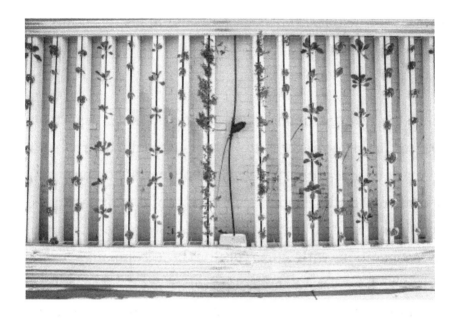

Hydroponics is swiftly expanding its momentum and reputation as the most vital and best method of cultivating every single kind of plant from food and flowers to medicinal herbs.

In Europe, a survey has shown that hydroponics is now generally trusted and affirmed by consumers. This means that hydroponically grown plants and foods are quickly catching on in several other countries and cities around the world. Farmers are learning new ways to cultivate healthier food crops and herbs. You should actually be well on your way to harvesting your first hydroponic crop by now. We sincerely hope that this book has answered all of your questions, and that I have given you a strong understanding of the hydroponic method. Your suggestions, opinions, and critiques about this book are sincerely welcomed so that I can make any necessary adjustments.

Since the hydroponic industry is still quite small, and there are not many local stores to buy supplies, the alternative is basically on-line garden stores that concentrate on providing hydroponic garden supplies or ready-to-use, prefabricated hydroponic gardens for people who cannot wait to get started. In cooperation with some of the best companies in the industry, however, we are endeavoring to provide a large selection of components, nutrients, and accessories that you may need to build, sustain, and maintain the gardens featured in this publication.

In the greenhouse, air circulation is very important for your gardening and can be supplied in two ways. The air circulation may be adequate if you can locate the greenhouse to take advantage of natural winds and have adequate ventilation openings. By using one or more fans in the greenhouse, ventilation can increase. If the temperature and humidity are allowed to spike during hot summer days, your gardening experience will not be very successful.

It's a good idea to install a ventilation system in your greenhouse. Plants can wilt under excess heat, so you're going to need a way to cycle cool air. This is an emergency measure and can delete any discoveries you have made, but it can save the plants, which is better than starting from scratch. A chest or cabinet that is waterproof is a perfect place to store your equipment in the greenhouse. This will protect them against rust damage and moisture while reducing the need to move in and out of the greenhouse—behavior that will risk the greenhouse's temperature.

Glossary

Aerating pump — All of God's creatures need oxygen, and fish are no exception. A little explanation of how your fish will use it illustrates the importance of supplying this basic life-sustaining element. Fish may not have lungs, but they still need air to breathe, filtering dissolved oxygen from the water through their gills.

The gills (usually four on each side) are situated just behind the head and are a series of bony flaps with filaments. Just as the human bronchus divides again and again to form a pulmonary tree for oxygen intake and exchange, the gill filaments support a network of lamellae. Water runs through the gills and those busy little filaments employ osmotic action to extract oxygen from the water with carbon dioxide being released in the exchange. That precious life-sustaining oxygen is directly absorbed into the fish's bloodstream.

Algae — Those nasty green strands in your tank are quite a nuisance, but worse, they can destroy your aquaponic garden. Algae are organisms growing in your water. Some are just single-celled organisms, and others grow in colonies creating chloroplasts as they grow. Left unchecked, they can cloud your water and make it unclean. Worse, algae affects the pH level and use precious oxygen in their growth.

Let's look at the oxygen issues first. Your fish need oxygen, and the last thing you want is to find them asphyxiated one morning due to oxygen depletion through the night. How does that happen, and what does algae have to do with it? Like all plants, algae will produce oxygen during daily photosynthesis. At night, when there is no light for photosynthesis, these strands will start to use your tank's reservoir of dissolved oxygen for their own growth, resulting in lower levels for your fish. If your fish appear to be suffering despite high oxygen levels, check your readings during the middle of the night. You may find them extremely low. Worse yet, when algae begins to die, the cells consume oxygen to decompose and further deplete your oxygen reservoir.

Bio filter — You'll hear this term a lot in aquaponic gardening, but it isn't complicated. A bio filter serves to extend your bed of microbes, converting fish waste to plant food. Yes, it's really that simple. It becomes an important consideration when you are submerging your plants into the water rather than cultivating them in a growth medium like hydroton pellets.

If you have a large operation with a raft of bedding plants and surface water to spare, you can utilize a floating medium like

K1 media to increase the surface area of microbes and provide adequate fertilization. You can also install static trays or drip filters and cycle your water through them between the solid waste filter and the submerged plants.

Your need for a bio filter is determined by the density of your fish population relative to the volume of water, as well as how many plants you want to grow. If you maintain a small number of fish and don't over feed them, you are going to be producing less solid waste and may be able to get by without a bio filter.

Deep water culture — This growing technique involves a deeper water supply with floating plants that rest on the surface, their roots submerged in the water. It alleviates the need for canals or tubes, pumping the water through a series of tanks, and is scalable from the tiniest of ambitions to large rafts of plants grown for commercial purposes.

Ebb and flow — This system of watering and feeding your plants is centuries old, and remains one of the simplest forms of aquaponics today. It is also known as a flood and drainage system. You will create your "flood table" by growing produce in a plastic tray with your growth medium positioned above the tank reservoir. Several times each day your sump pump is turned on to pump your watery stew up a pipe, emptying into your grow tray. Let it sit there and drain slowly until the next scheduled flooding/feeding for your plants. You can install a timer to handle the process automatically.

Fish food — Which food and how much of it? Nelson Pade offers a great organic, non-GMO food for fish, available in small or large bags online. You can also buy fish food from local pet stores or a nearby Walmart. What you feed fish is determined by the type of fish you're growing, but realize a few basic things. First, when you buy any kind of fish food, it will be loaded with a good mix of nutrients, a balance of protein, carbs, fats, and minerals.

Others opt for making their own. If you want to reduce costs, try duckweed, digging worms from the backyard, or slipping some larvae into your tank. If your fish are picky and won't eat these delicacies, you'll have to gather the remains to avoid excess strain on your filters.

Genetically modified organisms — Agribusinesses began experimenting with ways to improve on Mother Nature and devised ways of modifying the genetic structure of the food we eat. Their good intentions of reducing food shortages unleashed a storm of controversy over the benefits versus the harms of ingesting GMOs.

The rise in food-based allergies has risen from 3.4% in the late nineties to more than 5% in 2011. There is no evidence that genetically modified foods are responsible, but there is no evidence that it's not, either. The rising incidence of cancer worldwide, along with the increased development of superbugs, all provide fodder for the arguments against genetically modified foods. Again, there is no proof either way, but might I just suggest that it's not natural?

Grow lights — Who has enough sunlight for an indoor aquaponic garden? Not many of us. Luckily, you have lighting choices, and there are three main considerations. First, look at its size to be sure it will meet your garden's needs. Second, look at how it attaches. Some screw into existing light sockets and others are mounted to the ceiling. Third, look at the light's features.

If you need a larger system, look at the Philizon system of lighting, made in China. These range from a small version available on Amazon as a 600 LCD variety, up to multiple bars for extensive growing operations.

Herbs — These are great for aquaponic gardens because, unlike vegetables, you are growing small amounts to flavor dishes rather than whole meals.

Culinary herbs grow in many different ways. Some are perennials, some look like small shrubs, and others like small trees. Thyme, sage, lavender, parsley, basil, rosemary, and bay are the most commonly grown herbs. There are over four hundred medicinal herbs you can grow, but it requires knowing how to use the herb. Some are steeped into teas.

High tunnel — This is a structure built and covered with plastic. It serves as a mini greenhouse and can be either a simple protective covering or a fancy setup with installed watering systems, fans, and a heat source.

Homeostasis — Simply defined, it is the process of supporting life by maintaining a balance of neutrality within the body or organism. If you can understand the process in your own body, you have a frame of reference for understanding what is happening in your aquaponic garden. Your body employs many compensatory mechanisms for maintaining ideal blood sugar, blood pressure, body temperature, cellular water level, and pH, to name a few.

Your ideal body pH is 7.35 to 7.45. Within this narrow range, you are running your engines on all four cylinders. Suppose you start to hyperventilate. Your body will blow off carbon dioxide (CO_2), which is an acid, in each breath. That loss of CO_2 will make your blood more alkaline, lowering your blood ph. To compensate, your kidneys will kick in, excreting bicarb (HCO_3), which is the alkaline, and voila, your blood pH returns to a level within its normal range.

Hydroponics — Literally, this means growing plants in water. Many of the same principles are employed in an aquaponic garden.

Six types of hydroponic gardens may be constructed, and the first three require more ingenuity or trial and error than the latter three.

Hydration — These little gems are a form of dried clay broken up and baked into small pellets. If you've ever rooted plants in a glass of water, you know that the plant languishes against the side of the glass because its stem has no support of its own.

Most gardening centers offer bags of vermiculite and other soil-based media in large displays. Before entering the world of aquaponic gardening, I wasn't familiar with it either, but I soon learned that it offered five benefits. The little pellets are filled with tiny pores, and they are like sponges growing on the ocean floor. They absorb water but drain out excess moisture if you use them in a soil-based application. For our purposes, they hold water and provide surface area for necessary bacterial growth.

LECA aggregates — There are a number of options available for growing media. All of the pebbles in this category will work for you. You'll need small net pots that are 1.5 to 2.5 inches in diameter. Soak your pebbles and place seeds on top, covering them with one or two water-soaked pebbles, depending on recommended planting depth. Hydration is a brand of pellets, usually manufactured by heating clay in a rotating kiln. As it heats, the clay expands and forms pellets with incredibly increased surface areas for cultivating the bacteria you need as a growth medium.

Media bed — This system will look more like traditional gardening, since larger pots or trays hold the LECA (Lightweight Expanded Clay Aggregate) pellets and the plant grows looking like its traditional counterpart. The bed needs to be about 12 inches deep, and it increases the cost substantially.

The container you choose may be the most important decision you'll make. It will take up the largest amount of space, and it will need a depth best suited for the plants you want to grow. Fill it with pebbles, usually Hydration, to within 2 inches of the top. Remember that 12 inches is recommended. This growth bed acts as the bio filter.

If you are building a tower system with PVC pipes for your gardening bed, you're going to be investing in a lot more pellets. Don't skimp here. Less is not more; more is more.

Nitrogen cycle — This is the process sustaining life on planet earth. While nitrogen in its gaseous state comprises about 87% of our atmosphere, its life-sustaining qualities are not accessible by our bodies in its gaseous state.

Scientists call the first step in the cycle nitrogen fixation. Bacteria convert fish waste into ammonia. Ammonia (if you remember back to chemistry class) is chemically known NH3, and the process of nitrification begins.

Nutrient film — A nutrient film is a thin layer of nutrient-rich water that flows through plant roots for absorption. In hydroponic watering systems (which aquaponics is, merely replacing fish with chemicals), the water is distributed in channels, often without the aid of pumps. You'll read about wicking and ebb and flow in hydroponic gardening, but don't become overwhelmed. We eliminate the mystique in aquaponic gardening by using a small pump, which eliminates the need for a degree in physics, resulting in water being distributed automatically.

Organic gardening — Organic gardening is pure and unadulterated growth of plants without chemical fertilizers, herbicides, or pesticides. There have been many claims that these chemicals are linked to breast cancer, damaged brain function, Parkinson's disease, miscarriages, birth defects, autism, prostate cancer, non-Hodgkin's lymphoma, and infertility. Please realize a link is not definitive proof of causation, but it's enough to make a thinking person go, "Hmmm." There are other benefits as well. Many believe the compounded toxicity of constantly pouring chemicals into our soil, worldwide, will be catastrophic in years to come. Aquaponics offers a 100% organic gardening experience.

pH balance — Maintaining a neutral acid/base balance in your water is a matter of regulating the chemical reactions taking place. This is achieved, in part, by the nitrogen cycle, and nitrification forms the chemical basis for sustaining the aquatic and plant life in your aquaponic garden. Here's how the cycle is played out in your little microcosm of underwater drama. Your fish eat and convert their ingested proteins into ammonia (NH_3) and ammonium (NH_4+).

$$NH_3 + O_2 \rightarrow NO_2 + 3H+ + 2e-$$

Photosynthesis — You heard all about this in school, but it probably went in one ear and out the other. Now, it's time to get serious and be sure you understand what's involved. It is a chemical process in which plants take in sunlight and convert its energy into chloroplasts, keeping plants green and healthy. In this give and take reaction, water transfers electrons to carbon dioxide to produce carbohydrates. During the reaction, the carbon dioxide loses electrons and the water becomes oxidized. This cycle is Mother Nature's way of keeping all of the plants and animals happy through mutually beneficial trades.

Sump pump — Don't skimp on your pump. As you're counting the cost of building your aquaponic garden, there's always a temptation to channel your money into a fancier aquarium or into more fish. Resist that temptation. Your ecosystem will only be as good as its bones, and you need to invest in the right equipment.

Sustainability — We live in a delicate relationship with our home on Earth, and everything we do affects that relationship. When we grow and consume responsibly, nature and man live hand in hand. Our goal is to live with our environment– nurture it, not plunder it. For too many years, we have looked at the earth as a magical genie, stripping its natural resources with no thought of the consequences. Sustainability is a new way of looking at our earth home, and using our resources in a way that ensures future generations never get left with a bankrupt environment.

Symbiosis — in its most basic definition, symbiosis is the process of two organisms living together. In commensalism, one organism benefits from the other (think spider webs on trees). In parasitism, one lives off the other (think tapeworms). In mutualism, both organisms benefit, as in aquaponics. Your bacteria are tiny one-celled organisms you want to nurture and give a home to, because they are exchanging the ammonia your fish excrete into valuable fertilizer for your plants. These beneficial bacteria are not causing disease or going to infect you, so no worries. This is a win/win situation for the fish and for the plants. Find symbiotic worksheets to learn more!

CPSIA information can be obtained
at www.ICGtesting.com
Printed in the USA
BVHW041750011220
594609BV00005B/166